T0265372

A Theory of
Scattering
for Quasifree
Particles

A Theory of
Scattering
for Quasifree
Particles

Raymond F Streater

King's College London, UK

 **World Scientific**

NEW JERSEY • LONDON • SINGAPORE • BEIJING • SHANGHAI • HONG KONG • TAIPEI • CHENNAI

Published by

World Scientific Publishing Co. Pte. Ltd.

5 Toh Tuck Link, Singapore 596224

USA office: 27 Warren Street, Suite 401-402, Hackensack, NJ 07601

UK office: 57 Shelton Street, Covent Garden, London WC2H 9HE

Library of Congress Cataloging-in-Publication Data
Streater, R. F. (Raymond Frederick), author.
  A theory of scattering for quasifree particles / Raymond F. Streater, King's College London.
  pages cm
  Includes bibliographical references and index.
  ISBN 978-9814612067 (hardcover : alk. paper)
  1. Scattering (Physics) 2. Particles (Nuclear physics) I. Title.
  QC794.6.S3S77 2014
  539.7'58--dc23

                                2014011542

British Library Cataloguing-in-Publication Data
A catalogue record for this book is available from the British Library.

In-house Editor: Ng Kah Fee

Typeset by Stallion Press
Email: enquiries@stallionpress.com

Printed in Singapore

Contents

Chapter 1

Introduction

1.1 Introduction

> *If anyone could explain ... the peculiar properties*
> *of oxygen and hydrogen, ... he would have transferred*
> *certain phenomena from Chemistry into Physics. Those*
> *who have attempted to conceive of the chemical elements as*
> *different arrangements of particles of one primitive kind*
> *of matter, would, if they had been successful, have*
> *reduced Chemistry to Mechanics.*

Thus said J. C. Maxwell in his Inaugural Lecture, King's College London, 1860.

It is fair to say that physicists' attempts to describe the behaviour of elementary particles by relativistic quantum fields obeying the Wightman axioms [63] have not been a complete success. A similar statement is true for anyone trying to construct a theory obeying the Haag–Kastler axioms [36], which state physical properties that would hold for the C^*-algebra of local bounded observables. In any Wightman theory in a Hilbert space $\mathcal{H}$, with an isolated one-particle state of positive mass, the scattering states can be obtained by Haag–Ruelle theory [41] as strong limits of vectors in $\mathcal{H}$, and so also lie in $\mathcal{H}$. The same holds in a Haag–Kastler theory with an isolated one-particle state in the vacuum representation. This theory breaks down for models with particles of zero mass, such as quantum electrodynamics. It has been suggested [62] that the strong limit might be too strong, and that weak*-convergence is more appropriate. The use of a weaker form of convergence allows us to pass from one representation of the C^*-algebra to another, separated from the first by a superselection rule: we just need to follow a weak*-convergent sequence of states in one representation-space to its limit in another. Then it often turns out that convergence in the weak vector topology does in fact hold in the direct sum of the two representations.

It has been extremely difficult to construct interacting solutions to renormalisable quantum field theories that satisfy the Wightman axioms, in four space-time dimensions; it might be that only free and generalised free fields exist. The received wisdom of present-day quantum-field models of particle physics indicates that the interacting field becomes free in the limit of high energy. The starting point is an approximate theory, for example with some cut-offs at large energy, so that the modes with energy larger than the cut-off do not interact. The approximate answer for the S-operator is obtained by putting in some large cut-off at a deep level, and for a more accurate answer, one must use an even larger cut-off at an even deeper level [72]. We shall see one reason why the interaction may become zero at large energy in one of our models, that of the Higgs mechanism.

Albeverio and Hoegh-Krohn [3, 4] have generalised the Wightman axioms to a theory which allows a scalar product for which the norm is not always positive. Albeverio, Gottschalk and Wu [2] have given a class of quantum field theories obeying the Wightman axioms except the positivity of the metric. This gives rise to a finite theory with non-linear interaction; we proceed along different lines.

It has been conjectured that quantum electrodynamics does not exist; only theories containing quantised, non-abelian gauge vector fields, it is claimed, could exist as well as giving a non-trivial S-matrix. In this book we suggest a way round this question, using unusual representations of the C^*-algebra generated by the free transverse quantised electromagnetic field. Our work is inspired by some of Donaldson [29], and is applied to the cohomology of certain sheaves; we use the result that the differential structure of R^4, given by adding a non-trivial, flat cocycle to the free-field, is different from the usual representation in Fock space. The importance of Donaldson's result, that simply-connected four-space can have a different differential structure from that usually given to R^4, has also been suggested by Mickelsson [51]; in topological field theory, it appears in the work of Witten [73], and of Asselmeyer [10–14]; our suggestion is slightly different from theirs. In Penrose's book [53], as well as in Chapter 1 of Ref. 18, there is no discussion of the square-integrability of the solution, in the sense of whether or not it has finite norm as defined by Eq. (4.18). The paper [45] does impose square-integrability; thus these authors deal with the free theory. In the present book, we allow non-square-integrability; this gives rise to a flat connection on the sheaf of electromagnetic fields. This connection is a cocycle similar to the cocycles found by Doplicher and Roberts [30] in their analysis of superselection rules allowed by the Haag–Kastler axioms, and ensures the addition of a non-zero classical charge and current to the equations of motion. This leads us to our models given in Section 3 of Chapter 7.

The general theory of Haag fields has been analysed by Haag [30, 36, 37] and his coworkers, among them Kastler, Doplicher and Roberts. They obtain the result that the representation of the field might be reducible, having $SU(n)$ as the group of superselection rules, for any positive integer n. This result holds for any interacting quantum field theory obeying Haag's axioms. Now, in four dimensions, there is no known interacting quantum field theory: all known models are direct sums of free or quasifree quantum fields. However, these authors do not apply the theory to free fields, even though the results on the existence of superselection rules apply to these. We show that the theory is not trivial, when applied to the free electrodynamics of the photon field: there are quasifree representations of the photon field which possess charged states. More, there are quasifree representations in which particle–antiparticle pair has a non-zero overlap scalar product with a pair of photons: the transition probability of particle–antiparticle annihilation into a photon pair is positive.

Chapter 2

Haag–Kastler Fields

In four space-time dimensions, there is no known interacting model obeying the Wightman axioms, which were published long ago, in 1956. Haag and Kastler suggested their axioms later, but there is no interacting model known to obey them either. The relation between the Wightman axioms [63] and those of Haag and Kastler [37] is not clear for a general Wightman theory, but for any free boson field a key result due to Slawny [59] suggests a natural way to construct a set of local C*-algebras which obey the Haag–Kastler axioms. We briefly outline Slawny's construction, following Ref. [23]. Consider for example a free scalar quantised field $\hat{\phi}$ of mass $m > 0$. In any Lorentz frame, $\hat{\phi}$ and its time-derivative $\hat{\pi}$ at constant time (say, time zero) can each be smeared in the space variable with any function in the class C_0; that space comprises continuous functions $f, g, \ldots$ of compact support; Segal proved that we then get self-adjoint operators on Fock space. Thus

$$\hat{\phi}(g) := \int \hat{\phi}(0, \boldsymbol{x}) g(\boldsymbol{x}) d^3 x \qquad (2.1)$$

$$\hat{\pi}(f) := \int \hat{\pi}(0, \boldsymbol{x}) f(\boldsymbol{x}) d^3 x \qquad (2.2)$$

have well-defined imaginary exponentials; Segal proved that their sum is essentially self-adjoint on the domain of definition, and so also has a unique imaginary exponential, which is a bounded operator; let $\mathcal{K}_0$ be the space of real solutions of the Klein–Gordon equation such that the time-zero fields φ and ϖ lie in C_0. The space $\mathcal{K}_0$ has a pre-Hilbert structure, and its completion carries the representation $[m, 0]$ of the Poincaré group. The imaginary part of the scalar product in $\mathcal{K}_0$, given in Eq. (5.44), is the Wronskian of the

5

two solutions, denoted B by Segal:

$$B(\varphi_1, \varphi_2) := \int d^3x [\varphi_1(0, \boldsymbol{x})\varpi_2(0, \boldsymbol{x}) - \varpi_1(0, \boldsymbol{x})\varphi_2(0, \boldsymbol{x})]. \qquad (2.3)$$

This can be proved to be unchanged if both the solutions, φ_1 and φ_2 are subjected to a Poincaré transformation.

2.1   The Poincaré Group

The Poincaré group is the name given to the inhomogeneous Lorentz group, and was suggested by Bargmann, Wightman and Wigner [17]; these authors introduced the symbol $\mathcal{P}$ for this group, and we shall follow this. The (homogeneous) Lorentz group is the set $\mathcal{L}$ of real four-by-four matrices L that leave the metric tensor

$$\mathbf{g} = \begin{pmatrix} 1 & 0 & 0 & 0 \\ 0 & -1 & 0 & 0 \\ 0 & 0 & -1 & 0 \\ 0 & 0 & 0 & -1 \end{pmatrix} \qquad (2.4)$$

invariant under conjugation:

$$\mathcal{L} = \{L : LgL^t = g\}. \qquad (2.5)$$

In Eq. (2.5), L^t is the transpose of the matrix L. The product law is taken to be that of matrix multiplication. To see that the set $\mathcal{L}$ is a group, we first note that the unit 4×4 matrix obeys Eq. (2.5), and so lies in $\mathcal{L}$. Also, the product of two such transformations, L_1, L_2 is another such, and the inverse of any matrix obeying Eq. (2.5) always exists and also obeys Eq. (2.5). Thus the set of Lorentz matrices form a group with the group product taken to be matrix product.

In his theory of special relativity Einstein taught us that the laws of physics are invariant under the coordinate change

$$x \mapsto Lx \qquad (2.6)$$

where x is the four-vector of space-time, x_0, x_1, x_2, x_3, and L is any real matrix obeying Eq. (2.5); here of course the coordinate x_0 is equal to ct, where c is the speed of light and t is the time. One can show that for any $L \in \mathcal{L}$ we have $\det L = \pm 1$. However, in Nobel prize winning work, Lee and Yang [46] suggested that in weak interactions there might be a violation of invariance under parity, $S : \boldsymbol{x} \mapsto -\boldsymbol{x}$. This was almost immediately experimentally found to be the case by Wu and friends [74]. Most authors denote the parity operator by P; we use S, for *space*-inversion, as the symbol P denotes the four-vector of momentum, and also may be used for a general element of $\mathcal{P}$.

It turns out [28] that invariance under time-reversal, denoted by T, can also be violated, and the same goes for C, charge-conjugation [28]. Nevertheless, the product operation, SCT is always a symmetry in any relativistic local quantum field theory [63].

Space-time transformations which reverse parity, such as the transformation S:

$$t \mapsto t' = t$$

$$\boldsymbol{x} \mapsto \boldsymbol{x}' = -\boldsymbol{x}$$

lie in $\mathcal{L}$, but are *improper*; in fact, we write

$$S = \begin{pmatrix} 1 & 0 & 0 & 0 \\ 0 & -1 & 0 & 0 \\ 0 & 0 & -1 & 0 \\ 0 & 0 & 0 & -1 \end{pmatrix} \tag{2.7}$$

so that $S(ct, \boldsymbol{x}) = (ct, -\boldsymbol{x})$ and $\det S = -1$. The set of proper Lorentz matrices (those with $\det L = 1$) forms the subgroup denoted by $\mathcal{L}_+$. The set $\mathcal{L}_-$ consists of improper Lorentz matrices, those L with $\det L = -1$; these do not form a group, since the product of two elements of $\mathcal{L}_-$ lies in $\mathcal{L}_+$. The set of Lorentz matrices $\mathcal{L}$ can be written as the union of those that do not reverse the direction of time, $\mathcal{L}^{\uparrow}$, and those that do, $\mathcal{L}^{\downarrow}$. Thus:

$$\mathcal{L} = \mathcal{L}^{\uparrow} \cup \mathcal{L}^{\downarrow}. \tag{2.8}$$

Again, $\mathcal{L}^{\uparrow}$ is a group, while $\mathcal{L}^{\downarrow}$ is not. Since invariance under time-reversal might fail too [28], a Lorentz matrix which changes the sign of the time-coordinate might not be a symmetry of the system. We therefore adopt only the subgroup

$$\mathcal{L}_+^{\uparrow} = \{L \in \mathcal{L} : \det L = 1, L_{00} > 0\} \tag{2.9}$$

as a symmetry group of particle physics; this is a subgroup of $\mathcal{L}$, and is known as the group of *proper, orthochronous* Lorentz transformations. The Poincaré group, $\mathcal{P}$, includes the translations as well as Lorentz transformations:

$$x \mapsto x' = Lx + a \tag{2.10}$$

where $L \in \mathcal{L}$ and $a \in \mathbf{R}^4$. This is the semi-direct product of $\mathcal{L}$ and $\mathbf{R}^4$. Similarly, the groups $\mathcal{P}^\uparrow$ and $\mathcal{P}^\uparrow_+$ are defined as the semi-direct products of $\mathcal{L}^\uparrow$ and $\mathcal{L}^\uparrow_+$ with $\mathbf{R}^4$ respectively.

2.2   The Segal *-Algebra

For the free particle of spin zero and positive mass, Segal introduced the unitary operator, W, the exponential of i times the Wronskian $B(\hat{\phi}, \varphi)$ between the quantised field $\hat{\phi}(\boldsymbol{x}, t)$ and elements φ of the one-particle space $\mathcal{K}_1$ of real classical solutions:

$$W(\varphi) = \exp\left\{iB(\hat{\phi}, \varphi)\right\} = \exp\left(i \int d^3x \{\hat{\phi}\partial_t\varphi - \varphi\partial_t\hat{\phi}\}\right). \qquad (2.11)$$

Segal proved that $B(\hat{\phi}, \varphi)$ is essentially self-adjoint on $\mathcal{K}_0$. In terms of the Segal operators $W(\varphi), \varphi \in \mathcal{K}_1$, the Weyl relations Eq. (5.28) become the single relation

$$W(\varphi_1)W(\varphi_2) = \exp\left(-iB(\varphi_1, \varphi_2)/2\right)W(\varphi_1 + \varphi_2). \qquad (2.12)$$

This is Segal's form of the Weyl relations; it is discussed in more detail later; see Sec. 5.2. Segal's form gives a (non-commuting) product to the vector space over $\boldsymbol{C}$ generated by $W(\varphi)$ as φ runs over the symplectic space $\mathcal{K}_0$; one can multiply the elements by complex numbers, and add them, to get a *-algebra; one gets an isomorphic *-algebra, whatever the representation of the operators W. We call this the *Segal *-algebra*. What Slawny [59] did was to prove that the Segal *-algebra has a unique C*-norm; that is, a norm obeying

$$\|A^*A\| = \|A\|^2. \qquad (2.13)$$

We shall call this the *Slawny norm*.

Similar results can be proved for any free bosonic quantized field with positive mass and any spin.

We now define the Haag–Kastler field [8, 36, 37] for a system of Segal operators. Let $\mathcal{O}$ be a bounded open set in $\mathbf{R}^4$, of the form of the intersection of the interiors of a forward and a backward light cone. The surfaces of the cones themselves intersect in a two-dimensional set, which spans a three-dimensional region D inside both cones. Let f and g be continuous functions on D. Then the local C*-algebra $\mathcal{A}(\mathcal{O})$ is the completion, in the Slawny norm, of the Segal *-algebra generated by such f and g. The algebra of all observables, $\mathcal{A}$, is then the completion of the inductive limit of the union over $\mathcal{O}$ of all the local algebras; in this union, we of course identify the elements of $\mathcal{A}(\mathcal{O}_1)$ with a subalgebra of any $\mathcal{A}(\mathcal{O})$ whenever $\mathcal{O}_1 \subset \mathcal{O}$. That is, we form the union of all the algebras $\mathcal{A}(\mathcal{O})$, and then complete it in the Slawny norm. The algebra assigned to an arbitrary connected open region in $\mathbf{R}^4$ is the completion of the union of all $\mathcal{A}(\mathcal{O})$, $\mathcal{O}$ being a subset of the region. Note that this is not exactly the same set of operators as defined and used by Segal; in our algebra, every element is either localised in a bounded set $\mathcal{O}$, or, if it is obtained from the inductive limit, it is the norm limit of a sequence of localised elements. Segal adds the weak limits of some local elements, as well as including operators coming from all square-integrable solutions in his algebra.

2.3 The Haag–Kastler Axioms

The C*-algebra $\mathcal{A}$ defined by the previous paragraph nearly obeys the axioms of Haag and Kastler [37]; they assumed that the Poincaré group acted on $\mathcal{A}$ norm-continuously, which we do not. The free boson fields, and the observables of free fermion fields, satisfy one more axiom, the split property, (6) below, of Doplicher and Roberts [30]. Thus we shall postulate for a general theory:

1. A C*-algebra $\mathcal{A}$ is given, on which the Poincaré group acts as automorphisms.
2. To each bounded connected open set, $\mathcal{O} \subset \mathbf{R}^4$, is given a C*-algebra $\mathcal{A}(\mathcal{O})$; if a collection $\{\mathcal{O}_i\}$ of bounded open sets covers a given open set $\mathcal{O}$, then $\bigcup \mathcal{A}(\mathcal{O}_i)$ C*-algebraically generates a C*-algebra containing $\mathcal{A}(\mathcal{O})$.
3. The Poincaré group element (L, a) transforms $\mathcal{A}(\mathcal{O})$ onto $\mathcal{A}(L\mathcal{O} + a)$.
4. If two regions $\mathcal{O}_1$ and $\mathcal{O}_2$ are space-like separated, then the corresponding subalgebras, $\mathcal{A}_1$ and $\mathcal{A}_2$, commute.
5. The vacuum representation: there exists a representation R_0 of $\mathcal{A}$, which contains a unique vacuum vector, Ψ_0; that is, Ψ_0 is invariant under the Poincaré group, and is, up to a factor, the only vector with this property. Further, the Poincaré group is continuously represented by unitary operators, and the spectrum of energy is bounded from below.
6. The split property: If $\mathcal{O}_1^- \subset \mathcal{O}_2$ then there exists a subalgebra $\mathcal{N}$ of type I such that $\mathcal{A}(\mathcal{O}_1) \subset \mathcal{N} \subset \mathcal{A}(\mathcal{O}_2)$.

Here, the symbol $\subset$ means a proper subset, and type I means that the weak closure in the vacuum representation is a von Neumann algebra of type 1. Also, in (2), we say that a set of bounded operators, say $\mathcal{B}_0$, C*-algebraically generates an algebra $\mathcal{B}$, if and only if $\mathcal{B}$ is the intersection of all abstract C*-algebras containing $\mathcal{B}_0$.

In their set-up, Haag and Kastler [37] give the following explanation of superselection rules: charged states, and in fact any state that is separated by a superselection rule from the space of states containing the vacuum, are not in that Hilbert space, but are states in some other representation R of the algebra $\mathcal{A}$, which is not equivalent to R_0. They assume that the Poincaré group automorphisms are spatial in R, that is, they are implemented by unitary or anti-unitary operators in the Hilbert space on which R acts. They assume that the generator of time-evolution in the representation R has positive energy. R is related to the vacuum representation R_0 by an automorphism σ, $A \mapsto \sigma A$, $A \in \mathcal{A}$. The map σ cannot be a *spatial* automorphism, since if it were, the representations R and R_0 would be unitarily equivalent. Clearly, the representation R is given by

$$R(A) = R_0(\sigma A), \qquad (2.14)$$

as A runs over $\mathcal{A}$; the representation $A \mapsto R(A)$ acts on Fock space, but is not equivalent to the Fock representation, since the action of σ is not implemented by a unitary operator. Haag showed that one might reveal the existence of fermions, carrying a charge, by exploring the representations

$$R_n(A) = R_0(\sigma^n A), \qquad (2.15)$$

which would define the n-particle states.

Electromagnetism is described by the pair $z = (\boldsymbol{E}, \boldsymbol{H})$, defining the real pre-Hilbert space $\mathcal{K}$ with scalar product $\langle z_1, z_2 \rangle_{\mathcal{K}}$. $\mathcal{K}$ possesses a sympletic form, namely, the Wronskian $B(z_1, z_2)$. This furnishes $\mathcal{K}$ with a complex scalar product

$$\langle z_1, z_2 \rangle_C = \langle z_1, z_2 \rangle_{\mathcal{K}} + iB(z_1, z_2). \qquad (2.16)$$

The action of the Poincaré group is unitary, the representation being of mass zero and helicity 1. The transverse field obeys $\boldsymbol{E} \cdot \boldsymbol{H} = 0$ and $\boldsymbol{E} \cdot \boldsymbol{E} - \boldsymbol{H} \cdot \boldsymbol{H} = 0$. These expressions are Lorentz invariant. These operators allow us to define Segal's symbols $W(z)$ for each z, obeying the Segal relations

$$W(z_1)W(z_2) = W(z_1 + z_2) \exp\left\{ -\frac{i}{2} B(z_1, z_2) \right\}. \qquad (2.17)$$

The linear span over the complex field of the set of $W(z), z \in \mathcal{K}_L$ means the vector space of finite formal sums of the form $\Sigma \alpha_j z_j$; we can then use Eq. (2.17) to define the product, and the collection of these becomes an algebra. We define a *-operation on a single $W(z)$ as

$$W(z)^* := W(-z). \tag{2.18}$$

Finite sums are conjugated in the obvious way, starting with

$$(\alpha W(z))^* := \alpha^* W(-z). \tag{2.19}$$

The set of W generates a unique *-algebra, $\mathcal{A}$. We define the local operators associated with a bounded open set $\mathcal{O}$ in space at a given time, t, to be given by $W(z)$ where the support of $\boldsymbol{E}, \boldsymbol{B}$ associated with z lies in $\mathcal{O}$. Let $\mathcal{A}(\mathcal{O}, t)$ be the subalgebra of $\mathcal{A}$ localised in the region $\mathcal{O}$ at time t. According to Slawny, [59], $\mathcal{A}$ has a unique C*-norm. We complete the algebra $\mathcal{A}(\mathcal{O}, t)$ in this norm, for each time t and each bounded open set $\mathcal{O} \in \boldsymbol{R}^3$. We complete the union of all these algebras, for all time t and all bounded $\mathcal{O}$, in the norm topology, to get the algebra $\mathcal{A}$ of our theory. This algebra is invariant under the Poincaré group: the action of $\mathcal{P}$ on $\mathcal{A}(\mathcal{O})$, where $\mathcal{O}$ is a bounded open set of $\boldsymbol{R}^3$, on a time-slice at time zero, is an algebra in the bounded region $L\mathcal{O}$ on a space-like surface; this is a subalgebra of an algebra $\mathcal{A}(\mathcal{O}_1)$ defined at time zero, for a large enough but bounded region $\mathcal{O}_1$. Let R be the relativistic Fock representation of $\mathcal{A}$.

Given an automorphism σ of $\mathcal{A}$ we saw in Eq. (2.14) that we can get a new representation R of $\mathcal{A}$ from a given representation R_0 on a Hilbert space $\mathcal{H}$ by the definition

$$R(A) := R_0(\sigma(A)). \tag{2.20}$$

Doplicher and Roberts generalised this idea; this is well-explained in their article [30]. To get a representation of $\mathcal{A}$ one can make do with an endomorphism rather than an automorphism; in that case, one can get a reducible representation of the algebra $\mathcal{A}$. They limit the choice to endomorphisms, say σ, that preserve the identity:

$$\sigma(1) = 1. \tag{2.21}$$

The condition (2.21) ensures that the normalised vector states of the original representation R_0 are transformed into normalised vector states of the new representation, R. The unitaries of the commutant of the representation R make up the gauge group.

An *endomorphism* of a *-algebra $\mathcal{A}$ is a map $\sigma : \mathcal{A} \to \mathcal{A}$ such that

1. if $A \in \mathcal{A}$, then $\sigma(A^*) = (\sigma(A))^*$;
2. if $\lambda \in C$ and $A \in \mathcal{A}$, then $\sigma(\lambda A) = \lambda \sigma(A)$;
3. if A and B lie in $\mathcal{A}$, then $\sigma(A) + \sigma(B) = \sigma(A + B)$
4. if A and B lie in $\mathcal{A}$, then $\sigma(A)\sigma(B) = \sigma(AB)$.

There is no statement that σ has an inverse, or that $\sigma(\mathcal{A})$ should be the whole of $\mathcal{A}$.

This operator acts on $\mathcal{H}$ and obeys the rules of representation, in that (1), ... (4) lead to these for R. Thus, to prove property (1) for R,

$$
\begin{aligned}
R(A^*) &= R_0(\sigma(A^*))) \\
&= R_0((\sigma(A))^*) \\
&= (R_0(\sigma(A)))^* \\
&= (R(A))^*.
\end{aligned}
$$

To prove (2) for R, we get

$$
\begin{aligned}
R(\lambda A) &= R_0\{\sigma(\lambda A)\} \\
&= R_0\{\lambda \sigma(A)\} \\
&= \lambda R_0\{\sigma(A)\} \\
&= \lambda R(A).
\end{aligned}
$$

For (3),

$$
\begin{aligned}
R(A + B) &= R_0(\sigma(A + B)) \\
&= R_0(\sigma(A) + \sigma(B)) \\
&= R_0(\sigma(A)) + R_0(\sigma(B)) \\
&= R(A) + R(B).
\end{aligned}
$$

Finally, for (4),

$$
\begin{aligned}
R(AB) &= R_0(\sigma(AB)) \\
&= R_0(\sigma(A)\sigma(B)) \\
&= R_0(\sigma(A))R_0(\sigma(B)) \\
&= R(A)R(B).
\end{aligned}
$$

By definition, the endomorphisms constructed in this way obey the requirement $\sigma(1) = 1$.

Starting with axioms that are similar to our (1), ... ,(6), Haag, Doplicher and Roberts find [30, 36] that the gauge group must be a compact Lie group. Now, the axioms also apply to the algebra generated by the free field, though Doplicher and Roberts assumed that the given system was not the free field. They are stuck, in that no interacting Wightman theory in four dimensions has yet been constructed.

In this book, we start with the free quantised electromagnetic field, and show that there are some automorphisms and endomorphisms which do give rise to new states. In particular, for the free quantised electromagnetic field, there are automorphisms that lead to charged states. In these representations, the *transverse* components of the fields E, H obey the free wave equations of Maxwell. The longitudinal and time-like components of the fields, however, obey the interacting equations of Maxwell with non-zero charge and current. The charge and current are c-numbers, and in our theory, *should not* be quantised. The states in our new representations are weak*-limits of states in the Fock representation, obtained by scattering. We must include them to get an *asymptotically complete* theory.

Chapter 3

Representations
of the Poincaré Group

3.1 Induced Representations of Groups

Let G be a finite group, and let $H \subseteq G$ be a subgroup of G. Let V be a continuous unitary representation of H on a Hilbert space $\mathcal{E}$; let $\langle\,,\rangle_\mathcal{E}$ denote the scalar product on $\mathcal{E}$. Frobenius [32] showed how to construct a corresponding unitary representation U of G on a Hilbert space $\mathcal{K}$, which he called the representation *induced* by V. We shall write

$$U = V \uparrow_\mathcal{E}^\mathcal{K}. \tag{3.1}$$

The space $\mathcal{K}$ on which U acts consists of $\mathcal{E}$-valued maps on G, say ψ, that obey the equation

$$\psi(hg) = V(h)\psi(g) \tag{3.2}$$

for all $g \in G$ and $h \in H$; let $\mathcal{K}$ be the space of such maps. First, we note that $\mathcal{K}$ is a vector space over C; this follows at once from the fact that $\mathcal{E}$ is a vector space, and that V is a linear operator. Then the action of the representation $U(g)$, $g \in G$, induced by V is given by

$$(U(g)\psi)(g') = \psi(g'g). \tag{3.3}$$

This action of $U(g)$ maps $\mathcal{K}$ to itself; for, writing $\psi_g(g')$ for $\psi(g'g)$, we get from Eq. (3.2)

$$\psi_g(hg') = \psi(hg'g)$$
$$= V(h)\psi(g'g) \quad \text{(by Eq. (3.2))}$$
$$= V(h)\psi_g(g')$$

which is Eq. (3.2) for ψ_g at the point g'. The scalar product in the space G/H, given by

$$\langle \psi, \varphi \rangle := \sum_{g \in G} \langle \psi^*(g)\varphi(g) \rangle \varepsilon \qquad (3.4)$$

is then invariant under the action of $U(g)$, since V is unitary on $\mathcal{E}$.

If V is reducible, then so is U. However, it does not follow that if V is irreducible, then so is U. For example, if $V = 1$ on the one-dimensional Hilbert space C, then U is the regular representation of G; thus, the Hilbert space of the induced representation consists of functions on G, and the scalar product between two functions, ψ and φ is

$$\langle \psi, \varphi \rangle = \sum_{g \in G} \psi(g)^* \varphi(g). \qquad (3.5)$$

This representation is reducible unless G consists of only one element.

The construction of Frobenius can be generalised, for example, to the case where G is locally compact, and has a G-invariant measure $\mu(g)$, $g \in G$. Let H be a closed subgroup of G, continuously represented by unitary operators $V(h)$, $h \in H$, on the separable Hilbert space $\mathcal{E}$. The Hilbert space on which the induced representation acts uses $\mathcal{E}$-valued functions f on the group G that satisfy the identity

$$f(gh) = (V(h)f)(g) \qquad (3.6)$$

for all $h \in H$ and all $g \in G$. Equation (3.6) implies that the scalar product,

$$\langle f(hg), f(hg) \rangle = \langle V(h)f(g), V(h)f(g) \rangle = \langle f(g), f(g) \rangle \qquad (3.7)$$

is constant on each right coset Hg; it therefore defines a function, say $\|\tilde{f}\|^2$ on the right-coset space G/H; this space is furnished with the invariant measure, say $\tilde{\mu}$ coming from $\mu(g)$. The induced representation acts on the Hilbert space of almost-everywhere defined functions on G/H with finite norm, and for which

$$\int_{G/H} \|\tilde{f}\|^2 d\tilde{\mu} < \infty \qquad (3.8)$$

holds. The action of the induced representation U is given by right-multiplication:

$$(U(g)\tilde{f})(Hg') := \tilde{f}(Hg'g). \qquad (3.9)$$

This is unitary. For the proof of this, note that

$$\langle U(g)\tilde{f}, U(g)\tilde{f}' \rangle = \int_{G/H} d\tilde{\mu}(g')\langle f(Hg'g), f'(Hg'g) \rangle. \qquad (3.10)$$

We now replace the integration parameter g' by $g'g^{-1}$, to get

$$\langle U(g)\tilde{f}, U(g)\tilde{f}' \rangle = \int_{G/H} d\tilde{\mu}(g'g^{-1})\langle \tilde{f}(g'), \tilde{f}'(g') \rangle = \langle \tilde{f}, \tilde{f}' \rangle, \qquad (3.11)$$

since $\tilde{\mu}$ is invariant under the group G. Thus we have proved that $U(g)$ is unitary for all $g \in G$.

This theory can be further generalised to the case where the group G possesses a quasi-invariant, rather than an invariant measure; we omit the detailed discussion of this case.

The following theorem will be used later.

Theorem 1 *Let G be a group, and let H_1 and H_2 be subgroups of G, related by an inner automorphism $g_0 \in G$, thus:*

$$H_2 = g_0 H_1 g_0^{-1} \qquad (3.12)$$

for some $g_0 \in G$. Let V_1 be a representation of H_1 on Hilbert space $\mathcal{E}$, and let $U := V_1 \uparrow_{\mathcal{E}}^{\mathcal{K}}$ be the representation of G on the Hilbert space $\mathcal{K}$ induced by V_1. Then the map $V_2 : H_2 \to \mathcal{B}(\mathcal{E})$ given by

$$V_2(h_2) = V_1(g_0^{-1} h_2 g_0) \qquad (3.13)$$

is a representation of H_2 on $\mathcal{E}$, and it induces a representation of G that is equivalent to U.

For the proof, we see that the operators coming from V_2 are unitary, and that

$$\begin{aligned}
V_2(h_2)V_2(h_2') &= V_1(g_0^{-1}h_2g_0)V_1(g_0^{-1}h_2'g_0) \\
&= V_1((g_0^{-1}h_2g_0)(g_0^{-1}h_2'g_0)) \\
&= V_1(g_0^{-1}h_2h_2'g_0) \\
&= V_2(h_2h_2')
\end{aligned}$$

so we have proved that V_2 is a unitary representation of H_2 on $\mathcal{E}$. The representation of G induced by V_2 consists of functions f_2 on G that satisfy a square-integrability condition and the identity

$$f_2(gh_2) = (V_2(h_2)f_2)(g) \qquad (3.14)$$

for all $g \in G$ and all $h_2 \in H_2$. For each such function f_2 on G, define

$$f_1(g) = f_2(g_0 g g_0^{-1}). \qquad (3.15)$$

Then from Eq. (3.14) we see that

$$f_1(gh_1) = f_2(g_0 g h_1 g_0^{-1}) \qquad (3.16)$$

so we get

$$f_1(gh_1) = f_2(g_0 g g_0^{-1}(g_0 h_1 g_0^{-1}))$$
$$= f_2(g_1 h_2),$$

where

$$g_1 = g_0 g g_0^{-1}, \qquad (3.17)$$

giving

$$f_1(gh_1) = (V_2(h_2)f_2)(g_1)$$
$$= (V_1(h_1)f_1)(g). \qquad (3.18)$$

Thus the function f_1 satisfies Eq. (3.6), which is that required for right-multiplication to give the action of G in the representation U induced by V_1. Obviously, the same argument shows that the condition (3.18) leads to the condition (3.14), proving the theorem.

It was Wigner [71] who first extended the methods of induced representations to find representations of the Poincaré group. He found all the irreducible representations that represent particles occurring in physics. This work was made rigorous by G. W. Mackey, who also heroically generalised this theory [48] to other continuous groups. Mackey's work is designed to classify *all* continuous unitary projective representations of the group in question, as well as clarifying the details of the construction.

Wigner studied [71] the unitary and anti-unitary projective representations of the Poincaré group. By "projective" is meant that the group law holds on the states, rather than on the vectors of the Hilbert space; it holds only up to a complex factor of modulus one on the vector space. Recall that in elementary quantum mechanics, a state ψ is the collection of unit rays

$$\psi = \{\exp(i\alpha)\psi : \alpha \in \mathbf{R}\}. \qquad (3.19)$$

We shall limit our discussion to the space-time transformations without space or time reflection, and call the Poincaré group, $\mathcal{P}_+^\uparrow$, the group of transformations $\boldsymbol{R}^4 \to \boldsymbol{R}^4$ of the form

$$x \mapsto Lx + a, L \in \mathcal{L}_+^\uparrow, \quad a \in \boldsymbol{R}^4. \tag{3.20}$$

Thus $\mathcal{P}_+^\uparrow$ consists of the set of pairs $P = (L, a)$ obeying Eq. (3.20).

The product of two Poincaré transformations, $P_1 := (L_1, a_1)$ and $P_2 := (L_2, a_2)$ can easily be found; indeed, $P_1 x = L_1 x + a_1$, so

$$P_2 P_1 x = P_2 (L_1 x + a_1) \tag{3.21}$$

$$= L_2 (L_1 x + a_1) + a_2 \tag{3.22}$$

$$= L_2 L_1 x + L_2 a_1 + a_2. \tag{3.23}$$

Thus the multiplication law for $\mathcal{P}_+^\uparrow$ is

$$(L_2, a_2)(L_1, a_1) = (L_2 L_1, a_2 + L_2 a_1). \tag{3.24}$$

This has the form of the semi-direct product of the groups $\mathcal{L}_+^\uparrow$ and $\boldsymbol{R}^4$, with $\mathcal{L}_+^\uparrow$ acting on $\boldsymbol{R}^4$ according to Eq. (2.6). In Eq. (3.24), the element (L_1, a_1) acts first, and then (L_2, a_2) acts.

3.2 Wigner's Theory of Symmetry

According to Wigner, a quantum theory with a symmetry group G leads to a projective representation of the group by unitary or anti-unitary operators. More, if G acts measurably on the space of states, then there exist a measurable function

$$\omega(g, h), \quad G \times G \to U(1)$$

and unitary or anti-unitary operators $U(g)$ for all $g \in G$, such that

$$U(g)U(h) = \omega(g, h)U(gh) \tag{3.25}$$

for all $g, h \in G$. The function ω is called the multiplier. If $U(g)$ obeying Eq. (3.25) is given, the set of $U(g)$ is said to be a multiplier representation of G. If

$$\omega(g, h) = 1 \quad \text{for almost all } g, h \in G, \tag{3.26}$$

the representation is also said to be a true representation.

The possible functions $\omega(g, h)$ are measurable, but are ambiguous up to changes on a set of measure zero. That is, two multipliers that differ only on a set (in $G \times G$) of measure zero give the same set of ray-maps. They must satisfy $|\omega(g, h)| = 1$, for all g, h, with no exception, as can be seen by taking the modulus of Eq. (3.25). This is not the only relation that ω must satisfy; since for all $g, h, k \in G$ we must have

$$U(g)\{U(h)U(k)\} = \{U(g)U(h)\}U(k), \qquad (3.27)$$

we see that

$$U(g)U(hk)\omega(h, k) = U(gh)\omega(g, h)U(k) \qquad (3.28)$$

so that

$$U(g(hk))\omega(g, hk)\omega(h, k) = \omega(g, h)U((gh)k)\omega(gh, k). \qquad (3.29)$$

Now, the group G is associative, so that $g(hk) = (gh)k$ holds, and we can cancel $U(g(hk))$ with $U((gh)k)$ in Eq. (3.29), to get the *cocycle relation* obeyed by the multiplier:

$$\omega(g, h)\omega(gh, k) = \omega(g, hk)\omega(h, k) \qquad (3.30)$$

for all g, h and $k \in G$. If $g \mapsto U(g)$ is a true representation, so that $U(g)U(h) = U(gh)$ holds for all $g, h \in G$, we may consider also the representation

$$V(g) = \alpha(g)U(g) \qquad (3.31)$$

for some measurable function α. This will induce the same action as U on the observables of the theory, in that for any bounded operator $A \in \mathcal{B}(\mathcal{H})$ we have

$$U(g)AU(g)^{-1} = V(g)AV(g)^{-1} \qquad (3.32)$$

for all group elements $g \in G$. The V operators provide us with a projective representation, since they obey the equation

$$V(g)V(h) = \alpha(g)\alpha(h)U(g)U(h) \qquad (3.33)$$

$$= \alpha(g)\alpha(h)U(gh) \qquad (3.34)$$

$$= \alpha(g)\alpha(h)/\alpha(gh)V(gh) \qquad (3.35)$$

Thus, V is a multiplier representation, with $\omega(g, h)$ obeying the *coboundary condition*

$$\omega(g, h) = \alpha(g)\alpha(h)/\alpha(gh). \qquad (3.36)$$

We now check that such an ω does indeed obey the cocycle relation Eq. (3.30). Thus, let ω be a coboundary, so that it obeys Eq. (3.36). Then

$$\omega(g,h)\omega(gh,k) = \alpha(g)\alpha(h)/\alpha(gh) \times \alpha(gh)\alpha(k)/\alpha((gh)k)$$

so we get

$$\omega(g,h)\omega(gh,k) = \alpha(g)\alpha(h)\alpha(k)/\alpha(g(hk))$$
$$= \alpha(g)\alpha(hk)/\alpha(g(hk)) \times \alpha(h)\alpha(k)/\alpha(hk)$$
$$= \omega(g,hk)\omega(h,k),$$

and thus ω obeys the cocycle equation (3.30).

The set of measurable cocycles of a group G, where two cocycles differing on a set of measure zero are identified, forms an abelian group $\mathcal{Z}$ under multiplication, and the set of coboundaries is a subgroup, say $\mathcal{B}$. For, it is easy to check that if $\omega \in \mathcal{Z}$, then the inverse, $1/\omega$, is too; it is clear that the identity, $\omega(g) = 1$ for all $g \in G$, lies in $\mathcal{Z}$. Finally, if both $\omega_1 \in \mathcal{Z}$ and $\omega_2 \in \mathcal{Z}$ then their product $\omega = \omega_1\omega_2$ obeys, for all $g, h, k \in G$,

$$\omega(g,h)\omega(gh,k) = \omega_1(g,h)\omega_2(g,h)\omega_1(gh,k)\omega_2(gh,k)$$
$$= (\omega_1(g,h)\omega_1(gh,k))(\omega_2(g,h)\omega_2(gh,k))$$
$$= \omega_1(g,hk)\omega_1(h,k)\omega_2(g,hk)\omega_2(h,k)$$
$$= (\omega_1(g,hk)\omega_2(g,hk))(\omega_1(h,k)\omega_2(h,k))$$
$$= \omega(g,hk)\omega(h,k),$$

so $\omega \in \mathcal{Z}$ and $\mathcal{Z}$ is an abelian group. We see that the set of coboundaries $\mathcal{B}$ is a subgroup of $\mathcal{Z}$, and is invariant under conjugation by $\mathcal{Z}$ since the group is abelian:

$$h \mapsto ghg^{-1} = h \in \mathcal{B} \tag{3.37}$$

for all $g \in \mathcal{Z}$ and all $h \in \mathcal{B}$. The quotient,

$$\mathcal{C} := \mathcal{Z}/\mathcal{B} \tag{3.38}$$

is therefore a group, called the (first) cohomology group of G. It follows that if G is such that $\mathcal{C}$ contains only the identity, then for quantum theory with G as a symmetry group, the multiplier is a coboundary which can be removed from the representation by a choice of α in Eq. (3.31).

For the group $\mathcal{P}_+^\uparrow$, without the discrete elements of space- and time-reflection, every element is the square of another element. Now, the square of a unitary or anti-unitary operator is unitary; this implies that only unitary operators are needed to represent $\mathcal{P}_+^\uparrow$. The representation need not be irreducible, and might be projective. The latter holds in the case when particles of spin $1/2, 3/2, \ldots$ occur. Wigner proved the following result. Consider a projective representation $g \mapsto U(g)$ of the full group, $\mathcal{P}$, including reflections in space and time, and denote the multiplier by $\omega(g, h), g, h \in \mathcal{P}$. Restrict the representing unitary operators $U(g)$ to the subgroup of space-time translations $U(a), a \in \mathbf{R}^4$. Then Wigner proved that the multiplier for the translation subgroup, $\omega(a, b), a, b \in \mathbf{R}^4$ can be chosen to be the identity. This is in spite of the fact that the group $\mathbf{R}^4$ has got some non-trivial multipliers; they are not multipliers for the Lorentz group however. Thus, the restriction of the unitary-antiunitary projective representation of the relativity group to space-time translations is a unitary true representation, $a \mapsto U(a)$ of $\mathbf{R}^4$. This is, in general, reducible. It is well known that for an abelian group, all unitary irreducible true representations are one-dimensional, being of the form $U(a) = \exp\{ia^\mu p_\mu\}$. Here, $p_\mu, \mu = 0, 1, 2, 3$, lies in the dual space of space-time, and is the momentum of the state of the one-dimensional Hilbert space of the representation. Any unitary continuous true representation of an abelian group is the direct sum and integral of irreducible, and so of one-dimensional, representations.

It is physically observed that all energies of the electromagnetic field are positive. In an irreducible representation of $\mathcal{P}$, the squared mass $m^2 = p_\mu \mathbf{g}^{\mu\nu} p_\nu$ takes only one value (where $\mathbf{g}^{\mu\nu}$ is the metric tensor in Eq. (2.4)). For the classical electromagnetic field we need the mass to be zero. Thus the momentum $p = (p^0, \mathbf{p})$ of the classical electromagnetic field runs over the surface of the forward cone, except zero:

$$\mathcal{V}_0 := \{p \in \mathbf{R}^4 : p^0 > 0, \|\mathbf{p}\| = p^0\}. \tag{3.39}$$

The case $p = 0$ gives us zero electromagnetic field: it is the vacuum state.

Given $p \in \mathcal{V}_0$ and $p_0 > 0$, the subgroup of $\mathcal{L}_+^\uparrow$ that leaves p fixed is the set of rotations about the axis along $\mathbf{p}$. This is isomorphic to $SO(2)$; we use an angle $\theta \in [0, 2\pi)$ to parametrise this group, which acts on the translation subgroup $\mathbf{R}^4$. The *little group* defined by p is the semi-direct product of this group $SO(2)$ with the translation group $\mathbf{R}^4$. The (one-dimensional) representation V_s of the little group, acting on $\mathcal{E} = \mathbf{C}$, is given by multiplication:

$$V_s(\theta, a) = \exp\{is\theta\} \exp\{ip \cdot a\}. \tag{3.40}$$

This is unitary for each real value of s; it is of course irreducible, since $\dim \mathcal{E} = 1$. Wigner showed that to get a true representation, s must be an integer. To get a ray-map, s is also allowed to be a half-integer. The case $s = \pm 1$ gives us the representations of the photon, with left-handed and right-handed spinning particles. The case $s = \pm 1/2$ gives us the neutrino, which does not give rise to a true representation. In either case s is called the helicity of the states in the representation. The representation of $\mathcal{P}_+^\uparrow$, denoted by U_+ or $[0,1] = [m = 0, s = 1]$, is the representation induced by V_1:

$$U_+ := [m = 0, s = 1] = V_1 \uparrow_{\mathcal{E}}^{\mathcal{K}}. \tag{3.41}$$

In Eq. (3.41), $\mathcal{K}$ is the space on which U_+ acts.

It turns out from the theory of induced representations that we get the same representation, up to unitary equivalence, whatever the choice of $p \in \mathcal{V}_0$. Indeed, if we chose $q \in \mathcal{V}_0$ instead of p, then there is an element L of the Lorentz group such that $Lp = q$. Then by Theorem 1, the two representations of the group $G = \mathcal{P}_+^\uparrow$, the Poincaré group, using either little group, are unitarily equivalent.

The other representation of the little group, V_{-1} has helicity $s = -1$ in the formula (3.40), and induces the representation $U_- = [0, -1]$ of $\mathcal{P}_+^\uparrow$. The two representations U_+ and U_- are inequivalent representations of $\mathcal{P}_+^\uparrow$ and are connected by a space-reflection $S \in \mathcal{L}$, given by Eq. (2.7). The parity automorphism of $\mathcal{P}$ is then given by

$$(L, a_0, \boldsymbol{a}) \mapsto (SLS^{-1}, a_0, -\boldsymbol{a}). \tag{3.42}$$

The direct sum $U = U_+ \oplus U_-$ carries an irreducible representation of the *orthochronous* Poincaré group $\mathcal{P}^\uparrow$, containing space-reflections, S; the latter are represented by

$$U(S) = \begin{pmatrix} 0 & 1 \\ 1 & 0 \end{pmatrix}. \tag{3.43}$$

Time reversal needs an *anti-unitary* operator to implement it. This is shown in Theorem 3 to follow from the property that the energy is bounded below and unbounded above. One can add a time-reversal operator, say T, to the unitary operators $U(L, a), U(S)$ to get an unitary-antiunitary representation of the full Poincaré group: this group consists of all products of L, a, S and T in all orders. Free classical electromagnetism uses the representation of mass zero and helicity ± 1: the Hilbert space gives a quantum theory, the states of which are those of a single photon. This Hilbert space is called the *one-particle space*, denoted by $\mathcal{K}_1$ in this book.

3.3 Time-Reversal

Wigner [70] emphasised that the action of a symmetry group in quantum mechanics is not *given* as a unitary representation of the group on the Hilbert space of the theory. Rather, it is given as a set of probability-conserving transformations T of the space of states $\mathcal{K}$. In elementary quantum mechanics the space of states is the set of rays $\{\lambda\psi : \lambda \neq 0, 0 \neq \psi \in \mathcal{K}\}$; the transition probability from ψ to ϕ is $|\langle\phi,\psi\rangle|^2$ if both ϕ and ψ are normalised. Such a ray-map T is called a symmetry if it preserves the transition probabilities between states. Thus, a *symmetry* is a ray-map T such that

$$|\langle\phi',\psi'\rangle|^2 = |\langle\phi,\psi\rangle|^2 \qquad (3.44)$$

holds whenever $\phi \in \boldsymbol{\phi}$, $\psi \in \boldsymbol{\psi}$, $\phi' \in \boldsymbol{T\phi}$ and $\psi' \in \boldsymbol{T\psi}$ are unit vectors. It is clear that the set of symmetries forms a group, say G. Thus for each $g \in G$ we are given a map $\boldsymbol{U}(\mathrm{g}) : \mathcal{K} \to \mathcal{K}$ such that we have that

$$|\langle\boldsymbol{U}(\mathrm{g})\boldsymbol{\phi},\boldsymbol{U}(\mathrm{g})\boldsymbol{\psi}\rangle|^2 = |\langle\boldsymbol{\phi},\boldsymbol{\psi}\rangle|^2 \qquad (3.45)$$

holds for all $\boldsymbol{\phi}$ and $\boldsymbol{\psi}$ in $\mathcal{K}$. We then say that G is a symmetry group of the theory. In his book [70] Wigner gave us his result, known as Wigner's theorem:

Theorem 2 *Let $\boldsymbol{U}$ be a symmetry of a theory with Hilbert space $\mathcal{K}$; then there exists an operator $U : \mathcal{K} \to \mathcal{K}$, which is either unitary or anti-unitary, and is unique up to a complex factor of unit modulus, such that for any unit vector ϕ defining the state $\boldsymbol{\phi}$, we have that $U\phi$ defines the state $\boldsymbol{U\phi}$.*

We say that the symmetry $\boldsymbol{U}$ is implemented by U.

Wigner outlined a proof of this theorem in his book [70]; however, twenty-five years later, O'Raifeartaigh pointed out that Wigner's argument went nowhere near a proof. Bargmann rescued the situation by providing a beautiful and elementary proof, which however is quite long [16].

If G is a group of symmetries and is such that any element of G is the square of another element, then every symmetry $\boldsymbol{U}$ is implemented by a unitary rather than an anti-unitary operator; this is because the square of a unitary or anti-unitary operator is unitary. Thus, if G is a simply-connected Lie group, then all implementing operators are unitary.

We now come to an interesting point, which is to show that if a relativistic quantum theory obeys the axiom that the energy-spectrum is bounded below and is unbounded from above, and the dynamics is invariant under time-reversal, then time-reversal requires an anti-unitary operator to implement it. We prove this as our next theorem. We start with the fact, mentioned above, that every such theory leads to a continuous true representation of the space-time translation group. That is, no multipliers $\omega \neq 1$ need arise for the translation group. Then we have

Theorem 3 *Suppose that in a quantum theory on a separable Hilbert space with a continuous representation $U(t, \mathbf{a})$ of $\mathbf{R}^4$, obeying the axiom that the energy is bounded below but not above, and there exists a time-reversal operator T obeying*

$$TU(t, \mathbf{0}) = e^{i\alpha(t)}U(-t, \mathbf{0})T \qquad (3.46)$$

for some complex-valued function $e^{i\alpha}$ of unit modulus. Then T must be anti-unitary.

Proof The expectation of Eq. (3.46) in any state with energy in a bounded set shows that $\alpha(t)$ is C^∞ in a neighbourhood of $\alpha = 0$. The theorem is easily proved if the representation of $\mathbf{R}^4$ contains an irreducible representation other than the vacuum as a subrepresentation; for then this subrepresentation is one-dimensional, say $U(t, \mathbf{x})$; now put

$$U(t) = U(t, \mathbf{0}); \qquad (3.47)$$

then we have $U(t) = \exp\{ip_0t\}$, and

$$TU(t)T^{-1} = e^{i\alpha(t)}U(-t)$$

$$= e^{i\alpha(t)}\exp\{-ip_0t\}$$

for all real t. If T were unitary, then $U(t)$ with positive p_0 would be unitarily equivalent to $e^{i\alpha(t)}U(-t)$. Such a representation gives the same ray-map as $U(-t)$ with negative energy $-p_0$, which is untrue. Thus we have proved that in this case, T is anti-unitary.

Consider now the case when the representation $U(t)$ is a possibly reducible, multiplicity-free representation. For this, let μ be a positive measure on the dual space of the group, $X = \mathbf{R}^*$ and consider the Hilbert space of maps ψ, defined up to sets of measure zero, of μ-square-integrable functions from X to C, with scalar product

$$\langle \psi, \phi \rangle := \int_X \psi^*(p)\phi(p)d\mu(p). \tag{3.48}$$

The action of the group on this space is given by the unitary operator

$$(U(t)\psi)(p) := \exp\{ip_0 t\}\psi(p). \tag{3.49}$$

It is known that two such representations are unitarily equivalent if and only if the two measures are equivalent, that is, have the same sets of measure zero. The support of the measure is the energy-momentum spectrum of the model. We shall now assume that time-reversal is given by a unitary operator, and get a contradiction if the spectrum is non-negative and unbounded above. Indeed, the relation $U(t)T = e^{i\alpha(t)}TU(-t)$, acting on the state ψ, gives

$$\frac{\partial}{\partial t}U(t)T\psi = \frac{\partial}{\partial t}e^{i\alpha(t)}TU(-t)\psi. \tag{3.50}$$

This gives us at $t = 0$, if ψ lies in the domain of H:

$$T^{-1}HT\psi = -H\psi + \alpha'(0)\psi. \tag{3.51}$$

Now, if T is unitary, then the spectrum of $T^{-1}HT$ is the same as the spectrum of H, but Eq. (3.51) implies that this spectrum is that of $-H + \alpha'(0)$. Thus it contradicts the positivity, since the spectrum of $-H$ is unbounded below. Thus, by Wigner's theorem that a symmetry can be implemented by a unitary or anti-unitary operator, it must be possible to choose T to be anti-unitary.

Finally, we turn to the case, where $U(t)$ is a general representation of R on a separable Hilbert space. This is equivalent to [49]

$$U = U_\infty \oplus \bigoplus_{j \in S \subseteq N} U_j \tag{3.52}$$

where U_j is of multiplicity j and for $j \neq k$ the representations U_j and U_k are disjoint; in Eq. (3.52), N stands for the integers $\ldots, -2, -1, 0, 1, 2, \ldots$. Then U either contains an irreducible subrepresentation, or it contains a multiplicity-free subrepresentation. If T were unitary, then the restriction of T to one or the other subrepresentation would lead to the contradiction found above. We conclude from Wigner's theorem that it must be possible to choose T to be anti-unitary. This completes the proof of Theorem 3.

Chapter 4

The Maxwell Field

4.1 The Classical Electromagnetic Field

In this chapter, we shall denote the space of four-translations by a real four-vector a^μ, $\mu = 0, 1, 2, 3$, of space-time,

$$x^\mu \mapsto x^\mu + a^\mu \tag{4.1}$$

by $\boldsymbol{R}^4$; its dual space, momentum space, will be denoted by $\boldsymbol{R}^{*4}$. We shall use symbols involving p and k to denote momenta.

Let $\tilde{p} = (1, 0, 0, 1)$ be a four-vector on the forward light cone in $\boldsymbol{R}^{*4}$; then we define the little group of $\tilde{p}$, say $\mathcal{P}_0$, to be the subgroup of $\mathcal{P}_+^\uparrow$ which leaves $\tilde{p}$ invariant:

$$\mathcal{P}_0 = \{P \in \mathcal{P}_+^\uparrow : P\tilde{p} = \tilde{p}\}. \tag{4.2}$$

Then it is not hard to see that $\mathcal{P}_0$ is isomorphic to a semi-direct product of $SO(2)$ with $\boldsymbol{R}^4$, which we write as

$$H = SO(2) \triangle \boldsymbol{R}^4. \tag{4.3}$$

We denote a general element by (θ, a), where $0 \le \theta < 2\pi$, and $a \in \boldsymbol{R}^4$. We define the two one-dimensional representations

$$V_\pm(\theta, a) = \exp(\pm i\theta)\exp(i\tilde{p} \cdot a), \tag{4.4}$$

and form their direct sum, $V = V_+ \oplus V_-$. Consider the representation U of $\mathcal{P}_+^\uparrow$ induced from the representation V of $\mathcal{P}_0$ acting on $\mathcal{E} = \boldsymbol{C} \oplus \boldsymbol{C}$. The states on which U acts are then the $\mathcal{E}$-valued functions f on $G = \mathcal{P}_+^\uparrow$ obeying Eq. (3.6):

$$f(hg) = (V(\theta, a)f)(g) \tag{4.5}$$

when $h = (\theta, a)$. The action of $U(g)$ is by right-multiplication:

$$(U(g)f)(g') = f(g'g). \tag{4.6}$$

The set of functions f consists of those that are square integrable relative to the Lorentz-invariant measure on $\mathcal{P}_+^\uparrow$, modulo sets of zero measure.

The following is a modified version of part of Ref. 35. Let V_+ be the subset of $\boldsymbol{R}^{*4}$ consisting of momenta $k = (k_0, k_1, k_2, k_3)$ such that

$$k_0 > 0 \quad \text{and} \quad k \cdot k := k_0^2 - k_1^2 - k_2^2 - k_3^2 = 0. \tag{4.7}$$

Consider the complex-valued vector functions $\boldsymbol{e}(\boldsymbol{k})$ and $\boldsymbol{h}(\boldsymbol{k})$ defined on V_+ such that

$$\boldsymbol{k} \cdot \boldsymbol{e}(\boldsymbol{k}) = 0 \tag{4.8}$$

$$\boldsymbol{h} = \boldsymbol{k} \wedge \boldsymbol{e}(\boldsymbol{k})/k_0 \tag{4.9}$$

for each $k \in V_+$. Note that $dk_1\, dk_2\, dk_3/k_0$ is Lorentz-invariant. It follows from Eq. (4.9) that

$$\boldsymbol{k} \cdot \boldsymbol{h} = 0 \quad \text{and} \quad \boldsymbol{e} \cdot \boldsymbol{h} = 0 \tag{4.10}$$

as functions of $\boldsymbol{k}$. The one-particle space, containing photons of both helicities 1 and -1 is the set $\mathcal{K}_1$ of pairs $\boldsymbol{e}$, $\boldsymbol{h}$ of complex-valued vector functions on V_+ satisfying Eq. (4.8) and Eq. (4.9), and having finite norm:

$$\|(\boldsymbol{e}, \boldsymbol{h})\|^2 = \int_{V_+} \frac{\boldsymbol{e}(\boldsymbol{k}) \cdot \boldsymbol{e}(\boldsymbol{k})^* + \boldsymbol{h}(\boldsymbol{k}) \cdot \boldsymbol{h}(\boldsymbol{k})^*}{k_0^2} \frac{d^3 k}{k_0} < \infty. \tag{4.11}$$

To each element of $\mathcal{K}_1$ is associated the positive-energy solution of the free Maxwell equations,

$$\boldsymbol{E}(\boldsymbol{x}) = \int_{V_+} \boldsymbol{e}(\boldsymbol{k}) \exp\{i\boldsymbol{k} \cdot \boldsymbol{x}\} \frac{d^3 k}{k_0}, \tag{4.12}$$

$$\boldsymbol{H}(\boldsymbol{x}) = \int_{V_+} \boldsymbol{h}(\boldsymbol{k}) \exp\{i\boldsymbol{k} \cdot \boldsymbol{x}\} \frac{d^3 k}{k_0}. \tag{4.13}$$

This is complex-valued, and its time evolution has an analytic continuation into the upper-half time-plane. The norm can be written in terms of

E and H: let

$$f(k) = \frac{e}{k_0}. \tag{4.14}$$

Then at time zero, we have

$$E(x) = \int_{R^3} f(k) \exp\{ik \cdot x\} d^3 k \tag{4.15}$$

$$f(k) = (2\pi)^{-3} \int_{R^3} E(x) \exp\{-ik \cdot x\} d^3 x \tag{4.16}$$

The contribution of the electric field to the square of the norm in Eq. (4.11) is

$$\int \|f\|^2 \frac{d^3 k}{|k|}. \tag{4.17}$$

The contribution of h to the norm is the same as the contribution given by e. This is because the length of h is the same as the length of e, which follows from Eq. (4.9).

Gross shows that the norm in terms of E and H is

$$\|E, H\|^2 = \frac{4\pi}{(2\pi)^6} \int_{R^6} \frac{E(x) \cdot E(y)^* + H(x) \cdot H(y)^*}{|x - y|^2} d^3 x \, d^3 y. \tag{4.18}$$

The fields E and H have no longitudinal components, so they obey div $E = 0$ as well as div $H = 0$; they are also mutually orthogonal, so $E \cdot H = 0$; finally, in the units we use, they have the same length: $E \cdot E = H \cdot H$.

The Hilbert space $\mathcal{K}_1$ is the space of the one-photon states. The classical field description is given by the real parts of E, H, which contains states of negative energy, as well as positive energy. We denote the real part of E, H by E_c and H_c. In the second-quantised form, the states of negative energy correspond to the annihilation of the photon, while the states of positive energy correspond to the creation of the photon.

We prove in Theorem 4 that a smooth real classical electromagnetic field E_c, H_c, equal to that derived from the potential

$$V = e/|x| \tag{4.19}$$

outside a ball, moving at constant velocity, is not square integrable, in that its norm is divergent to infinity. Then by Shale's theorem [57], Theorem 6, it gives rise to a non-Fock representation when we add this classical field to the usual free quantised field, $\widehat{\boldsymbol{E}}_0$, $\widehat{\boldsymbol{H}}_0$, thus: let

$$(\widehat{\boldsymbol{E}}, \widehat{\boldsymbol{H}}) = (\widehat{\boldsymbol{E}}_0, \widehat{\boldsymbol{H}}_0) + (\boldsymbol{E}_c, \boldsymbol{H}_c). \tag{4.20}$$

Indeed, we have

Theorem 4 *We get a vector of infinite norm when we represent the quantised transverse electromagnetic field by adding a classical field $F_{\mu\nu}$ caused by a non-zero charge e at the origin $\boldsymbol{x} = \boldsymbol{0}$, moving with a constant velocity.*

Proof If the charge is not moving, then it creates a classical field

$$\boldsymbol{E}_c = -e\partial_{\boldsymbol{x}}|\boldsymbol{x}|^{-1} = e|\boldsymbol{x}|^{-2}\hat{\boldsymbol{x}}; \tag{4.21}$$

the value of $\boldsymbol{H}$ is zero. This term contributes to the norm of the classical field $\boldsymbol{E}_c$, $\boldsymbol{H}_c$ a term proportional to the infinite integral given in Eq. (4.18) with $\boldsymbol{H} = 0$ and

$$\boldsymbol{E} = \boldsymbol{E}_c. \tag{4.22}$$

Then we get for the contribution to the norm of a region outside a ball of radius L, an expression proportional to

$$\iint \frac{\boldsymbol{E}_c(\boldsymbol{x}) \cdot \boldsymbol{E}_c(\boldsymbol{y})^*}{|\boldsymbol{x} - \boldsymbol{y}|^2} d^3x \, d^3y$$

$$= \int_L^\infty r dr \, d\Omega \int_L^\infty r' dr' \, d\Omega' F(\Omega, \Omega') r^{-2} r'^{-2} |\boldsymbol{x} - \boldsymbol{y}|^{-2}.$$

for some non-zero function F of the angular variables. Here, $r = |\boldsymbol{x}|$ and $r' = |\boldsymbol{y}|$. Now, $|\boldsymbol{x} - \boldsymbol{y}|^2 \leq (r + r')^2$, so its inverse is greater than or equal to $(r + r')^{-2}$. Then the integral over r and r' is greater than or equal to

$$\int^\infty dr \int^\infty dr'(r + r')^{-2} \tag{4.23}$$

which diverges at ∞. Thus the norm of $\boldsymbol{E}_c$ is infinite, and we get a non-Fock representation by choosing the representation $\boldsymbol{E} = \boldsymbol{E}_0 + \boldsymbol{E}_c$.

We can allow the source of the potential to be a smooth function of $r = |\boldsymbol{x}|$ of compact support inside the sphere $\{\boldsymbol{x} : r < L\}$ with total charge q. This is because such a distribution gives rise to the potential q/r outside its support. We shall see that q must be an integer multiple of a minimal value, which we call e. This will follow from the requirement that the connection whose differential is being discussed should be one-valued in the space $\boldsymbol{R}^3$. We can transform this state by a Lorentz transformation, and get a moving charge and a current. However, we cannot get a two-particle representation if we split the charge e into two parts, and apply a different Lorentz transformation to each part, such as sending one part towards negative x and the other towards positive x. For then as time goes by, the two parts of the field separate, but the state does not converge, as time passes, into two particles of half the charge; this is because the sum of two fields, each of charge $e/2$, gives a state, but do not give a state of two particles if separated by taking the limit to infinite time, when the connection becomes many-valued. This completes the proof of Theorem 4.

In this representation, space translation is represented by unitary operators; for, the difference between the representation given in Eq. (4.20) and its space translation is a function that is square integrable.

4.2   The C*-norm for Electromagnetism

In this section we show, using the theorem of Slawny, that there is a unique C*-algebra for the transverse quantum electromagnetic field obeying the Maxwell equations with any classical charge-current. For non-zero charge, the representation of the C*-algebra differs from the representation with zero charge.

The covariance of $\boldsymbol{E}$, $\boldsymbol{H}$ as a tensor follows from the fact that $\boldsymbol{e}$, $\boldsymbol{h}$ obey the Eqs. (4.8), (4.9). These lead to the relativistic free Maxwell equations for $\boldsymbol{E}$, $\boldsymbol{H}$. That is, it follows that the matrix

$$F^{\mu\nu} := \begin{pmatrix} 0 & -E_1 & -E_2 & -E_3 \\ E_1 & 0 & H_3 & -H_2 \\ E_2 & -H_3 & 0 & H_1 \\ E_3 & H_2 & -H_1 & 0 \end{pmatrix} \tag{4.24}$$

transforms as a second-order tensor under $\mathcal{L}_+^\uparrow$.

In this theory, we allow a flat connection on the sheaf of electromagnetic fields. This connection is a cocycle similar to the cocycles found by Doplicher and Roberts [30] in their analysis of superselection rules allowed by the Haag–Kastler axioms, and leads to the addition of a non-zero classical charge and current to the equations of motion.

In classical Maxwell theory, as well as in quantum field theory, the tensor $F_{\mu\nu}$, can be written in terms of a four-vector potential, which for the free Maxwell equations consists of four functions on space-time, A^μ, $\mu = 0, 1, 2, 3$, such that

$$F^{\mu\nu} = \partial^\mu A^\nu - \partial^\nu A^\mu. \tag{4.25}$$

The potential is not observable, and is not unique. Indeed, we may change A^μ by a *gauge transformation*

$$A^\mu \mapsto A^\mu + \partial^\mu \Phi \qquad (4.26)$$

for any differentiable function Φ of space-time, without changing $F^{\mu\nu}$.

For the free classical electromagnetic field, the expression

$$\nabla_\mu := \partial_\mu - ieA_\mu \qquad (4.27)$$

is said to be an affine connection on the space of smooth anti-symmetric second-order tensor functions depending on time and space. These make up a *tensor bundle*. Here, affine means that it behaves linearly on convex mixtures of sections of the bundle of smooth real tensors over space R^3 of compact support. In point of fact, ∇_μ is a *differential*, to be applied to the section of the bundle; the connection itself is then got by integrating this between the two points of the manifold which we are connecting together. If $A_\mu = 0$, this is clear, as then $\nabla_\mu = \partial_\mu$, and we may integrate $\partial_j F$ round a closed curve $\mathcal{C}$. This gives

$$\oint_{\mathcal{C}} i\partial_j F\, dx_j = 2\pi i(F(x_2) - F(x_1)) = 0 \qquad (4.28)$$

since for a closed curve $\mathcal{C}$, the start point x_1 equals the final point x_2. Thus the term involving ∂ contributes the factor $\exp(0) = 1$ to the expression

$$\exp i \left(\oint_{\mathcal{C}} \nabla \cdot d\boldsymbol{x} \right) \qquad (4.29)$$

when acting on the tensor bundle. At a fixed time, say $t = 0$, and when $A^0 = 0$, we have a three-vector $\boldsymbol{A}^j$, $j = 1, 2, 3$, leading to the three-vector ∇^j. This must define a connection in R^3. This means that the connection from a point $\boldsymbol{x}$ to itself, along any smooth path, must be the identity. If this does not hold, the connection is many-valued: it would define a connection only locally on R^3.

In order to be able to replace ∂ by ∇ as the differential of a connection, we see that, for any closed smooth curve $\mathcal{C}$, we have

$$\exp i \oint_{\mathcal{C}} \nabla \cdot d\boldsymbol{x} = 1 \qquad (4.30)$$

when acting on the tensor bundle. Now, the operator, say $S = \oint_{\mathcal{C}} \partial \cdot d\boldsymbol{x}$, commutes with $T = \oint_{\mathcal{C}} \boldsymbol{A} \cdot d\boldsymbol{x}$, which is independent of $\boldsymbol{x}$. Thus we have

$$\exp\left\{ i \oint (\partial - ie\boldsymbol{A}) \cdot d\boldsymbol{x} \right\} = \exp\left\{ i \oint \partial \cdot d\boldsymbol{x} \right\} \exp\{-ie\boldsymbol{A} \cdot d\boldsymbol{x}\}. \qquad (4.31)$$

Apply this operator to any constant field F, which must be unchanged by Eq. (4.30); we see that

$$\exp\left\{ i \oint \partial \cdot d\boldsymbol{x} \right\} F = F, \tag{4.32}$$

so we are led to

$$\exp\left\{ -ie \oint \mathbf{A} \cdot d\boldsymbol{x} \right\} F = F. \tag{4.33}$$

Thus we are left with the condition of quantisation for the field $\boldsymbol{A}$:

$$\oint_C e\boldsymbol{A} \cdot d\boldsymbol{x} = 2n\pi \tag{4.34}$$

where n is an integer. This condition is obviously sufficient for F to be invariant. The case $n = 1$ will then represent the smallest charge for any particle in the theory.

The representation of the observable algebra using Eq. (4.20) is inequivalent to the Fock representation when $n = 1$; however, the C*-algebra is the SAME as the free one: we have changed the representation, but kept the same algebra, that of the transverse modes.

We can also use a more general representation of the observable algebra, obtained from the Fock representation by an endomorphism. Then the identity on the right-hand side of Eq. (4.30) is replaced by the unit matrix in $SU(2)$, $SU(3)$, ..., depending on the dimension of the matrix occurring in the endomorphism used. For the fields $\boldsymbol{E}$, $\boldsymbol{H}$ to be functions of space and time, the integral of ∇ round any closed curve must be the identity. When expressed in terms of A_μ, the integral of ∇ round any closed curve in $\boldsymbol{R}^4$ must be a gauge transformation.

Let us first analyse the condition (3.6), in the special case when G is the Poincaré group $\mathcal{P}_+^\uparrow$, and the group element g is the identity, I, so that $(\Lambda, a) = (1, 0)$. Then Eq. (3.6) says

$$f(h) = (V_1(\theta, a)f)(I), \tag{4.35}$$

which gives us the specific behaviour for the translation subgroup which f must obey:

$$f(1, a) = \exp(i\tilde{p} \cdot a)f(I). \tag{4.36}$$

Thus, the non-trivial dependence of f on the group element is its dependence on the Lorentz component, Λ. The little group can be taken as the semi-direct product K of the group $\boldsymbol{R}^4$ acted on by the group

$$SO(2) = \{\Lambda \in L_+^\uparrow : \Lambda\tilde{p} = \tilde{p}\}. \tag{4.37}$$

Then the manifold $\mathcal{M}$ on which f is a function is

$$\mathcal{M} = \mathcal{P}_+^\uparrow / K = \mathcal{L}_+^\uparrow / SO(2). \tag{4.38}$$

Penrose [53] showed that the smooth *transverse* free electromagnetic field is given by a holomorphic or anti-holomorphic function on $\boldsymbol{PC}^3$, the complex projective space in three variables. This space is a compact, complex analytic manifold [68]. In that book, as well as in Chapter 1 of Ref. 18, there is no discussion of the square integrability of the solution, in the sense of whether or not it has finite norm as defined by Eq. (4.18). The paper [45] does impose square integrability; thus these authors deal with the free theory. In the present book, we allow non-square-integrability; this gives rise to a flat connection on the sheaf of electromagnetic fields. This connection is a cocycle similar to the cocycles found by Doplicher and Roberts [30] in their analysis of superselection rules allowed by the Haag–Kastler axioms, and ensures the addition of a non-zero classical charge and current to the equations of motion. This leads us to the models given in Section 3 of Chapter 7.

Chapter 5

Some Theory of Representations

5.1 The Tensor Product and Fock Space

Let $\mathcal{H}_1$ and $\mathcal{H}_2$ be two complex separable Hilbert spaces, which could be the state-spaces of two quantum systems. The tensor product

$$\mathcal{H}_1 \otimes \mathcal{H}_2 \tag{5.1}$$

will denote the Hilbert space-state of the quantum system containing both. In this book, this notation is reserved for the *algebraic* tensor product of the Hilbert spaces $\mathcal{H}_1$ and $\mathcal{H}_2$. We shall use $\mathcal{H}_1 \overline{\otimes} \mathcal{H}_2$ for the completion of the algebraic tensor product, which is a separable Hilbert space.

More generally, the algebraic tensor product of two vector spaces $\mathcal{H}_1$ and $\mathcal{H}_2$ is defined as follows: the set of pairs $(\Phi_1, \Phi_2) \in \mathcal{H}_1 \times \mathcal{H}_2$ generated by the vector space of finite formal sums of its elements with complex coefficients. Let us denote this vector space by Σ. If $\lambda \in C$ is not equal to 1, the elements $\lambda(\Phi_1, \Phi_2)$ and $(\lambda \Phi_1, \Phi_2)$ are different elements of Σ, whereas we need $\lambda(\Phi_1 \otimes \Phi_2)$ to be equal to $(\lambda \Phi_1) \otimes \Phi_2$ for all $\lambda \in C$. The set M of elements in Σ of the form

$$M = \{\sigma(\lambda) : \lambda \in C\} \tag{5.2}$$

$$= \{\lambda(\Phi_1, \Phi_2) - (\lambda \Phi_1 \Phi_2) : \lambda \in C\} \tag{5.3}$$

forms an additive vector subspace of Σ: indeed we have that in Σ, $\lambda = 0$ is the zero of the subspace, and any two elements $\sigma(\lambda)$ and $\sigma(\lambda')$ can be added in Σ to get the element $\sigma(\lambda + \lambda')$. Let us denote this subspace by $M(\Phi_1, \Phi_2)$; it is also an invariant subgroup, since addition in Σ is commutative. The quotient space $\Sigma/M(\Phi_1, \Phi_2)$ treats $M(\Phi_1, \Phi_2)$ as zero. The element $\lambda(\Phi_1, \Phi_2)$ should also be regarded as equal to $(\Phi_1, \lambda \Phi_2)$, leading us to introduce the subgroup $N(\Phi_1, \Phi_2)$ of Σ given by elements of the form

$$N(\Phi_1, \Phi_2) = \{\lambda(\Phi_1, \Phi_2) - (\Phi_1, \lambda\Phi_2) : \lambda \in \boldsymbol{C}\}. \tag{5.4}$$

This, like M, is an invariant subgroup, and so is the group, say $Q = M \times N$; then the quotient group, Σ/Q treats the elements of Q to be zero. In this way, we define the subspace Σ_0 of Σ to be the vector space generated by the following:

$$\lambda(\Phi_1, \Phi_2) - (\lambda\Phi_1, \Phi_2) \tag{5.5}$$

$$\lambda(\Phi_1, \Phi_2) - (\Phi_2, \lambda\Phi_2) \tag{5.6}$$

$$(\Phi_1 + \Psi_1, \Psi_2) - (\Phi_1, \Psi_2) - (\Psi_1, \Psi_2) \tag{5.7}$$

$$(\Phi_1, \Phi_2 + \Psi_2) - (\Phi_1, \Phi_2) - (\Phi_1, \Psi_2) \tag{5.8}$$

as λ runs over $\boldsymbol{C}$, Φ_1, Ψ_1 run over $\mathcal{H}_1$ and Φ_2, Ψ_2 run over $\mathcal{H}_2$. From abelianness, Σ_0 is an invariant subgroup of Σ. The quotient space Σ/Σ_0 thus defines a vector space, denoted here by $\mathcal{V}$, called the algebraic tensor product of the vector spaces $\mathcal{H}_1$ and $\mathcal{H}_2$. The elements of Σ given by Eqs. (5.5)–(5.8) are zero in $\mathcal{V}$. We denote the equivalence class in Σ containing the element (Φ, Ψ) by $\Phi \otimes \Psi$. Then in $\mathcal{V}$ we get the relations corresponding to the Eqs. (5.5)–(5.8):

$$\lambda(\Phi_1 \otimes \Phi_2) = (\lambda\Phi_1) \otimes \Phi_2 \tag{5.9}$$

$$\lambda(\Phi_1 \otimes \Phi_2) = \Phi_1 \otimes (\lambda\Phi_2) \tag{5.10}$$

$$(\Phi_1 + \Psi_1) \otimes \Psi_2 = \Phi_1 \otimes \Psi_2 + \Psi_1 \otimes \Psi_2 \tag{5.11}$$

$$\Phi_1 \otimes (\Phi_2 + \Psi_2) = \Phi_1 \otimes \Phi_2 + \Phi_1 \otimes \Psi_2. \tag{5.12}$$

We use the symbol $\otimes$ for this product: $\mathcal{H} = \mathcal{H}_1 \otimes \mathcal{H}_2$.

In quantum theory, there is a scalar product on the vector spaces used. So suppose that $\mathcal{H}_1$ and $\mathcal{H}_2$ have scalar products $_1\langle\ ,\ \rangle_1$ and $_2\langle\ ,\ \rangle_2$. We define the putative scalar product on $\mathcal{H}$ to be

$$\langle(\Phi_1, \Phi_2), (\Psi_1, \Psi_2)\rangle = {}_1\langle\Phi_1, \Psi_1\rangle_1 \times {}_2\langle\Phi_2, \Psi_2\rangle_2. \tag{5.13}$$

This vanishes on the elements of Σ of the form of any of the equations (5.5) to (5.8), and so defines a scalar product on Σ/Σ_0. Thus, the algebraic tensor product of two scalar-product spaces is a scalar-product space. The algebraic tensor product of two Hilbert spaces is thus a scalar-product space, but it might not be complete if the factors have infinitely many dimensions. Its completion can easily be shown to be a scalar-product space obeying the relations (5.9)–(5.12). We shall use the symbol $\overline{\otimes}$ for the completion of the algebraic tensor product of two Hilbert spaces, $\mathcal{H}_1 \otimes \mathcal{H}_2$. Rays in this space are then the states of the combined quantum systems.

Similarly, states of three, four, ... systems are given by the spaces of three-fold, four-fold, ... tensor products. There is a natural isomorphism between $\mathcal{H}_1 \overline{\otimes}(\mathcal{H}_2 \overline{\otimes} \mathcal{H}_3)$ and $(\mathcal{H}_1 \overline{\otimes} \mathcal{H}_2) \overline{\otimes} \mathcal{H}_3$; these can be identified, so we can leave out brackets in multiple tensor products. However, when the systems are particles and $\mathcal{H}_1$ and $\mathcal{H}_2$ are the same, such as systems of photons say, then Bose made a suggestion that was equivalent [22] to saying that photons are such that only the symmetric sums

$$(\Phi_1 \otimes \Phi_2)_S := 2^{-\frac{1}{2}}[\Phi_1 \otimes \Phi_2 + \Phi_2 \otimes \Phi_1] \tag{5.14}$$

$$\cdots$$

$$\left(\bigotimes_{j=1}^{n} \Phi_j\right)_S := (n!)^{-\frac{1}{2}} \sum_{\sigma \in S(n)} \Phi_{\sigma(1)} \otimes \ldots \otimes \Phi_{\sigma(n)} \tag{5.15}$$

$$\cdots$$

occur in the statistical mechanics of photons. In Eq. (5.15), $S(n)$ is the permutation group on n elements $1, 2, \ldots, n$. The general boson state of n identical particles is the symmetrised state, and such states span the space $(\otimes_{j=1}^{n} \mathcal{K})_S$.

Let $\mathcal{K}_1$ be the state-space of a single particle. The boson Fock space over $\mathcal{K}_1$, called $\mathcal{K}$ in this book, is then defined as the completion of

$$\mathcal{K}_0 = C \oplus \mathcal{K}_1 \oplus \cdots \oplus (\mathcal{K}_1 \otimes \cdots \otimes \mathcal{K}_1)_S \oplus \cdots \tag{5.16}$$

$$= C \oplus \mathcal{K}_1 \oplus \bigoplus_{i=2}^{\infty} (\mathcal{K}_1 \otimes \cdots \otimes \mathcal{K}_1)_S, \tag{5.17}$$

where in Eq. (5.17) the tensor product $\otimes \cdots \otimes$ occurs i times. This space contains states with n bosons for any non-negative integer n.

The *Fock representation* of the free boson field, $\hat{\varphi}$ with test-functions Φ taken from the Hilbert space $\mathcal{K}_1$, is determined by the action on vectors in Fock space:

$$\hat{\varphi}(\Phi)\{\Phi_1 \otimes \cdots \otimes \Phi_n\}_S := \{\Phi \otimes \Phi_1 \otimes \cdots \otimes \Phi_n\}_S. \tag{5.18}$$

In the following, we shall use the brackets $\langle$ and $\rangle$ for the scalar product in both $\mathcal{K}_1$ and in $\mathcal{H}$.

We now come to exponential vectors, which are closely related to coherent states. Let Φ be a vector in a Hilbert space $\mathcal{K}_1$, taken as the one-particle space. Define [7] the vector in Fock space, written as e^{Φ}, by

$$\exp \Phi := e^{\Phi} := 1 \oplus \Phi \oplus 2^{-\frac{1}{2}} \Phi \otimes \Phi \oplus (3!)^{-\frac{1}{2}} \Phi \otimes \Phi \otimes \Phi \oplus \cdots. \qquad (5.19)$$

Clearly, we see that for all vectors $\Phi, \Psi \in \mathcal{K}_1$ we have

$$\langle e^{\Phi}, e^{\Psi} \rangle = \exp\langle \Phi, \Psi \rangle. \qquad (5.20)$$

We now show that the set of exponential vectors $\{e^{\Phi} : \Phi \in \mathcal{K}_1\}$ is total in Fock space; that is, we show that finite linear combinations of the exponential vectors make up a dense subset. Our proof of this result, using the method of induction on the number of factors n, is due to Araki [7]. For $n = 0$, $e^0 = 1 \in C$, and this vector spans $C = (\otimes \mathcal{K})^0$, so the theorem is true for $n = 0$. Now assume each $(\overline{\otimes}\mathcal{K})_S^j$, $j = 0, 1, \ldots, n-1$ is in the closure of the space spanned by the vectors $\{e^{\Phi}\}$, $\Phi \in \mathcal{K}$. Consider the vector, for $\lambda \in C$,

$$\Phi_n(\lambda) = e^{\lambda \Phi} - \Sigma_{j=0}^{n-1}(j!)^{-1/2}(\otimes \lambda \Phi)^j. \qquad (5.21)$$

Then it is easy to show that

$$\lim_{\lambda \to 0} \Phi_n(\lambda)/\lambda^n = (n!)^{-1/2}(\otimes \Phi)^n \qquad (5.22)$$

in the sense of strong convergence. Since by the induction hypothesis the right-hand side of Eq. (5.21) lies in the strong closure of the span of the exponential vectors, we have proved the induction hypothesis for n, and hence the claimed result.

The following result shows the boson nature of the construction.

Theorem 5 *There is a Hilbert-space isomorphism implemented by a unitary operator U between $\exp(\mathcal{H}_1 \oplus \mathcal{H}_2)$ and $\exp(\mathcal{H}_1) \otimes \exp(\mathcal{H}_2)$ such that*

$$U \exp(\Phi_1 \oplus \Phi_2) = \exp(\Phi_1) \otimes \exp(\Phi_2). \qquad (5.23)$$

Proof For any $\Phi = \Phi_1 \oplus \Phi_2 \in \mathcal{H}_1 \oplus \mathcal{H}_2$ we have

$$\exp\{\|\Phi\|^2\} = \exp\{\|\Phi_1\|^2 + \|\Phi_2\|^2\}.$$

So we have

$$U(\Phi) = \exp\{\|\Phi_1\|^2\} \cdot \exp\{\|\Phi_2\|^2\}$$

$$= \|\exp\{\Phi_1\} \otimes \exp\{\Phi_2\}\|^2.$$

Now the tensor product of the vectors in two total sets make up a total set in the tensor product. So the operator U defined above is unitary on a total set, and so can be extended to a unitary operator. This proves the theorem.

5.2   Segal's Form of the Weyl Relations

Elementary books on quantum mechanics start with the Heisenberg commutation relations between the operators q and p taken to represent the position and momentum of a particle in one dimensional space:

$$qp - pq = \hbar i. \tag{5.24}$$

If the book goes on to say, as many do, that Schrödinger found the unique irreducible representation of this equation, up to unitary equivalence, then it would be wrong. It is true to say that on the space of square integrable functions, the choice

$$q = \text{multiplication by } x$$

$$p = -i\hbar \frac{d}{dx} \tag{5.25}$$

does satisfy Eq. (5.24) on a dense set of vectors. However, these operators are not bounded, and there are many domains of functions on which the Heisenberg relations hold. Weyl transformed the problem from that of a non-compact Lie algebra with the relation (5.24) to the corresponding Lie group. He used two continuous unitary groups $U(\alpha) : \alpha \in \mathbf{R}, V(\beta) : \beta \in \mathbf{R}$, which correspond to self-adjoint generators p and q:

$$U(\alpha) = \exp\{i\alpha q\} \tag{5.26}$$

$$V(\beta) = \exp\{i\beta p\}. \tag{5.27}$$

These are required to satisfy the *Weyl relations*:

$$U(\alpha_1)U(\alpha_2) = U(\alpha_1 + \alpha_2)$$

$$V(\beta_1)V(\beta_2) = V(\beta_1 + \beta_2)$$

$$U(\alpha)V(\beta) = \exp\{-i\hbar\alpha\beta\}V(\beta)U(\alpha) \tag{5.28}$$

for all real numbers $\alpha_1, \beta_1, \alpha_2, \beta_2, \alpha, \beta$. The Stone–von Neumann theorem states that any unitary operators $\{U(\alpha), V(\beta) : \alpha, \beta \in \mathbf{R}\}$ on a separable

Hilbert space and continuous in α, β, is unitarily equivalent to a direct sum
of copies of the usual representation on $L^2(\boldsymbol{R})$ given by

$$\{U(\alpha)\psi\}(x) = \exp\{i\alpha x\}\psi(x) \qquad (5.29)$$

$$\{V(\beta)\psi\}(x) = \psi(x + \hbar\beta). \qquad (5.30)$$

The generators, x and $p := -i\partial_x/\hbar$, of the unitary groups $U(\alpha)$ and $V(\beta)$
then obey Eq. (5.24) on a well-defined domain. Thus, subject to the require-
ment of separability, there is only one way to set up an irreducible quantum
theory with one degree of freedom.

For a system with n degrees of freedom, the analogue of the Stone–von
Neumann theorem gives us the same conclusion: provided that n is finite,
there is, up to unitary equivalence, only one irreducible representation of
the corresponding Weyl relations. However, this uniqueness fails for a system with infinitely many degrees of freedom. Thus, uniqueness fails for a
system of one or more quantum fields. Irving Segal developed this idea in
a useful way. As background, let us give the idea for the case of the free
Hermitian quantum scalar field of mass $m > 0$. The field acts on the Fock
space, of which the one-particle space, $\mathcal{K}_1$, is a complex Hilbert space carrying the representation $[m, 0]_+$ of the Poincaré group. The suffix $_+$ restricts
the representation to positive energies. Here we must decide on whether
the quantised field which describes the particle is to be regarded as observable, or if not, whether it is charged. If it is to be observable, we choose a
real solution, denoted by $\varphi(t, \boldsymbol{x})$ say. However, then it does not appear to
lie in a complex Hilbert space, and moreover, its space-time Fourier transform has its spectrum in the backward light cone as well as the forward
light cone, and so might seem to fail the spectrum condition. Choosing a
complex, positive-energy solution, say $\phi(t, \boldsymbol{x})$, allows us to define $\mathcal{K}_1$ to consist of the positive-energy complex solutions $\phi(t, \boldsymbol{x})$ of the Klein–Gordon
equation

$$(\partial_t^2 - \partial_x^2 - \partial_y^2 - \partial_z^2)\phi(t, x, y, z) + m^2\phi(t, x, y, z) = 0. \qquad (5.31)$$

Here, $\boldsymbol{x} = (x, y, z)$. However, we now show that the required real solution
φ is a transform of this case, with a non-local relation between $\varphi(t, \boldsymbol{x})$
and $\phi(t, \boldsymbol{x})$. An element of the space $\mathcal{K}_1$ is determined for example by the
pair of functions of $\boldsymbol{x}$, ϕ and $\partial_t\phi$ at time zero. The set of solutions, for
which both functions $\phi(\boldsymbol{x})$ and $\partial_t\phi(\boldsymbol{x})$ lie in the complex Schwartz space
$\mathcal{S}(\boldsymbol{R}^3)$, is dense in $\mathcal{K}_1$. The requirement of positive energy means that such
a solution is analytic in the upper-half-plane in t, the time variable; it

follows that it is not possible to choose the time-dependence of its real and imaginary parts independently. More, the fact that $\phi(t, \boldsymbol{x})$ is a complex function means that it is not itself immediately an observable. It is sometimes used to describe a free charged particle; we shall not use this idea. In the theory to be presented in this book, charged particles are never free.

For any mass $m > 0$, a realisation of the representation $[m, 0]_+$ of $\mathcal{P}_+^\uparrow$ is to choose the Hilbert space of complex functions

$$\mathcal{K} = L^2(\boldsymbol{R}^3, \omega^{-1}d^3p), \tag{5.32}$$

where $\omega := (\boldsymbol{p} \cdot \boldsymbol{p} + m^2)^{1/2}$, and to define for all p on the forward light cone the unitary operator $U_0(\Lambda, a)$, by

$$(U_0(\Lambda, a)\tilde{\phi})(\boldsymbol{p}) = \exp\{-i(\omega t - \boldsymbol{p} \cdot \boldsymbol{x})\tilde{\phi}((\Lambda^{-1}p)_s). \tag{5.33}$$

In Eq. (5.33), when q is a four-vector, then $q_s = \boldsymbol{q}$ denotes the space-part of q.

Let us consider this case further. To each $\tilde{\phi} \in \mathcal{K}$ we can define a complex field $\phi(t, \boldsymbol{x})$ in space-time by

$$\phi(t, \boldsymbol{x}) = \int \omega^{-1}d^3p \ \exp(-i\omega t + i\boldsymbol{x} \cdot \boldsymbol{p})\tilde{\phi}(\boldsymbol{p}). \tag{5.34}$$

This field is complex-valued, satisfies the Klein–Gordon equation (5.31), and its space-time inverse Fourier transform is

$$\tilde{\phi}(p^0, \boldsymbol{p}) = \delta(p \cdot p - m^2)\theta(p^0)\tilde{\phi}(\boldsymbol{p}). \tag{5.35}$$

In this equation, $p \cdot p$ is the Lorentz-invariant scalar product $p_0^2 - \boldsymbol{p}^2$. The presence of the step-function $\theta(p^0)$ in the formula guarantees that the energy of all states is non-negative.

There is another realisation of the representation $[m, 0]_+$, unitarily equivalent to this one, which is closer to the canonical formalism of classical mechanics, in which the field is real, and so could correspond to an observable when quantised. Let $\mathcal{M}$ denote the set of *real* solutions φ to the Klein–Gordon equation (5.31) which, at time $t = 0$ have Cauchy data in the Schwartz space $\mathcal{D}(\boldsymbol{R}^3)$:

$$\varphi(0, \boldsymbol{x}) = f(\boldsymbol{x})$$

$$\partial_0\varphi(0, \boldsymbol{x}) = g(\boldsymbol{x}).$$

It can be proved that the pair (f, g) uniquely determine $\varphi(x_0, \boldsymbol{x})$ by Green's formula

$$
\begin{aligned}
\varphi(x_0, \boldsymbol{x}) = \int_{y_0=0} d^3 y \{ & \varphi(0, \boldsymbol{y}) \partial_0 \Delta(x - y; m) \\
& - \partial_0 \varphi(0, \boldsymbol{y}) \Delta(x - y; m) \},
\end{aligned}
$$

where Δ is the Green function obeying the Klein–Gordon equation, and at time zero, the equations

$$
\Delta(0, \boldsymbol{x}) = 0
$$

$$
\partial_0 \Delta(0, \boldsymbol{x}) = \delta^3(\boldsymbol{x})
$$

hold. The set of real solutions to the Klein–Gordon equation forms a Hamiltonian system, with canonical variables $\varphi(0, \boldsymbol{x})$ and $\varpi(0, \boldsymbol{x}) = \partial_t \varphi(0, \boldsymbol{x})$ at time zero; the Hamiltonian is

$$
\mathcal{H} = \frac{1}{2} \int d^3 x [m^2 \varphi^2(0, \boldsymbol{x}) + \varpi^2(0, \boldsymbol{x}) + \nabla \varphi(0, \boldsymbol{x}) \cdot \nabla \varphi(0, \boldsymbol{x})]. \tag{5.36}
$$

Green's formula can be written

$$
\begin{aligned}
\varphi(t, \boldsymbol{x}) = \frac{1}{2(2\pi)^{3/2}} \int \{ & \tilde{\phi}(\boldsymbol{k}) \exp[i\omega(k)t - i\boldsymbol{k} \cdot \boldsymbol{x}] \\
& + \tilde{\phi}^*(\boldsymbol{k}) \exp[-i\omega(k)t + i\boldsymbol{k} \cdot \boldsymbol{x}] \} \frac{d^3 k}{\omega(k)}, \tag{5.37}
\end{aligned}
$$

where in Eq. (5.37), $\omega(k) = (k^2 + m^2)^{1/2}$, and $\tilde{\phi}(\boldsymbol{k})$ is related to f, g by

$$
f(\boldsymbol{x}) = \frac{1}{2(2\pi)^{3/2}} \int \{ \tilde{\phi}(\boldsymbol{k}) + \tilde{\phi}^*(-\boldsymbol{k}) \} \exp\{-i\boldsymbol{x} \cdot \boldsymbol{k}\} \frac{d^3 k}{\omega(k)}
$$

$$
g(\boldsymbol{x}) = \frac{i}{2(2\pi)^{3/2}} \int \{ \tilde{\phi}(\boldsymbol{k}) - \tilde{\phi}^*(-\boldsymbol{k}) \} \exp\{-i\boldsymbol{k} \cdot \boldsymbol{x}\} d^3 k.
$$

The inverse of this map is

$$
\tilde{\phi}(\boldsymbol{k}) = \frac{1}{(2\pi)^{3/2}} \int (\omega(k) f(\boldsymbol{x}) - ig(\boldsymbol{x})) \exp(i\boldsymbol{k} \cdot \boldsymbol{x}) d^3 x. \tag{5.38}
$$

The presence of the non-local operator ω in Eq. (5.38) shows that the real field $\varphi(t, \boldsymbol{x})$ and the complex field $\phi(t, \boldsymbol{x})$ are non-local relative to each other; for example, if $\varphi(0, \boldsymbol{x})$ has compact support in a set $\mathcal{O} \subset \boldsymbol{R}^3$, then

$\phi(0, \boldsymbol{x})$ will be non-zero for most of the complementary set to $\mathcal{O}$. These fields are related by

$$\varphi(t, \boldsymbol{x}) = \operatorname{Re} \phi(t, \boldsymbol{x}) \tag{5.39}$$

$$\partial_t \varphi(t, \boldsymbol{x}) = \operatorname{Re} \partial_t \phi(t, \boldsymbol{x}). \tag{5.40}$$

The imaginary part of the scalar product between two elements of $\mathcal{K}_1$ is the Poisson bracket for the pair of solutions. For the real scalar field φ, this is the Wronskian

$$\{\varphi_1, \varphi_2\} = \int d^3x (\varphi_1(0, \boldsymbol{x}) \varpi_2(0, \boldsymbol{x}) - \varphi_2(0, \boldsymbol{x}) \varpi_1(0, \boldsymbol{x})), \tag{5.41}$$

where $\varpi_i(t, \boldsymbol{x}) = \partial_t \varphi_i(t, \boldsymbol{x})$ for $i = 1, 2$. This expression is of course invariant under the action of $\mathcal{P}$ on the scalar field; indeed, the $\mathcal{P}$-invariant scalar product on $\mathcal{K}_1$ is

$$\langle \varphi_1, \varphi_2 \rangle = \int_{p_0 > 0} d^4p \, \delta(p_0^2 - \|\boldsymbol{p}\|^2 - m^2) \tilde{\phi}_1(p)^* \tilde{\phi}_2(p)$$

$$= \int \frac{d^3p}{2\omega(\boldsymbol{p})} \tilde{\phi}_1(\boldsymbol{p})^* \tilde{\phi}_2(\boldsymbol{p}). \tag{5.42}$$

This can be expressed in terms of the space-integral of the corresponding real solution, $\varphi(t, \boldsymbol{x})$ of the Klein–Gordon equation with mass m, and its time-derivative $\partial \varphi / \partial t = \varpi(t, \boldsymbol{x})$ at any time, say $t = 0$:

$$\langle \varphi_1, \varphi_2 \rangle = \frac{1}{2} \int_{\boldsymbol{R}^3} d^3x (\omega^{\frac{1}{2}} \varphi_1(0, \boldsymbol{x}) \omega^{\frac{1}{2}} \varphi_2(0, \boldsymbol{x})$$

$$+ \omega^{-\frac{1}{2}} \varpi_1(0, \boldsymbol{x}) \omega^{-\frac{1}{2}} \varpi_2(0, \boldsymbol{x}))$$

$$+ \frac{i}{2} \int_{\boldsymbol{R}^3} d^3x (\varphi_1(0, \boldsymbol{x}) \varpi_2(0, \boldsymbol{x}) - \varpi_1(0, \boldsymbol{x}) \varphi_2(0, \boldsymbol{x})). \tag{5.43}$$

The symbol ω is the pseudo-differential operator $(-\Delta + m^2)^{\frac{1}{2}}$. This is a non-local operator, so the real part of the scalar product is non-local. The imaginary part is, however, local in the space-time variable $(t, \boldsymbol{x})$.

We shall need the following theorem, proved by Shale [57]:

Theorem 6 *Let* $\varphi \mapsto W_0(\varphi)$ *be the Fock representation of the relativistic free quantised field of mass* $m \geq 0$ *(containing the vacuum state). Let* T *be a symplectic operator on* $\mathcal{K}_1$ *and* Φ *be any real c-number* *solution to the Klein–Gordon equation. Then the representation R of the algebra $\mathcal{A}_S$*

$$R(W(\varphi)) = W_0(T\varphi + \Phi) \tag{5.44}$$

is unitarily equivalent to the Fock representation if and only if both

(1) $1 - |T|$ is of Hilbert–Schmidt class;
(2) $\Phi \in \mathcal{K}_1$

hold.

In this theorem, the algebra $\mathcal{A}_S$ is Segal's version of the C*-algebra generated by the canonical operators $W(\varphi)$ as φ runs over the one-particle space. This is slightly bigger than the algebra used by Haag and Kastler, in that Segal's algebra contains some weak limits. It also contains elements that are not norm limits of wave functions localized in a bounded set in $\boldsymbol{R}^4$, as long as the wave function lies in $\mathcal{K}_1$. In (1), $|T| = (T^*T)^{\frac{1}{2}} \geq 0$.

Recall that $\mathcal{K}_1$ is the Hilbert space of square integrable solutions of the free Klein–Gordan equation. Segal regarded $\mathcal{K}_1$ as a real vector space, with symplectic form B equal to the imaginary part of the scalar product; a symplectic map T is then a real linear map leaving B invariant: $B(\varphi_1, \varphi_2) = B(T\varphi_1, T\varphi_2)$ for all $\varphi_1, \varphi_2 \in \mathcal{K}_1$.

The case when

$$T = 1 \tag{5.45}$$

leads to a representation by *coherent states* determined by the solution Φ.

For the Maxwell equations for a transverse electromagnetic field, a similar result holds; one replaces φ, the Hermitian solution of the Klein–Gordon equation, by a transverse solution $\boldsymbol{E}$, $\boldsymbol{H}$ of the free Maxwell equations, ((5.47)–(5.50)), and Φ becomes a real transverse solution of the interacting Maxwell equations with real-valued charge density ρ and current density j_μ.

Let us now follow the same ideas for the mass-zero, spin 1 representation U_T of the Poincaré group,

$$U_T := [m = 0, s = 1] \oplus [m = 0, s = -1] \tag{5.46}$$

related to the transverse electromagnetic field described by the free Maxwell equations, obeyed by the pair E, H:

$$\text{div}\, E = 0 \tag{5.47}$$

$$\text{div}\, H = 0 \tag{5.48}$$

$$\partial_t\, B = -\text{curl}\, E \tag{5.49}$$

$$\partial_t\, E = \text{curl}\, B. \tag{5.50}$$

The expression, $B(\varphi_1, \varphi_2) = \{\varphi_1, \varphi_2\}$ vanishes if the two solutions φ_1 and φ_2 of the Klein–Gordon equation (or the Maxwell equations) and their time derivatives have supports that are disjoint at say time $t = 0$. Then the operators $W(\varphi_1)$ and $W(\varphi_2)$ obey the law $W(\varphi_1)W(\varphi_2) = W(\varphi_1 + \varphi_2) = W(\varphi_2)W(\varphi_1)$. This expresses relativistic causality, in that it leads to the proof that any two operators of the theory constructed from z_1 and z_2 in two space-like separated regions of space-time, commute with each other. Thus, the corresponding operators can be simultaneously measured. This is the principle of causality.

Segal and his students defined the C*-algebra, $\mathcal{A}_S$ of the theory, to be the following. For any finite number of classical solutions, say $\varphi_1, \ldots, \varphi_n$, by the Stone–von Neumann theorem there is a unique irreducible representation of the C*-algebra generated by elements $\{W(\varphi_j) : j = 1, 2, \ldots, n\}$, up to unitary equivalence. They therefore uniquely define the C*-algebra as the weak closure of this algebra in any one of them. Segal identifies elements of two such algebras if they represent the same element $W(\varphi)$, and then takes the inductive C*-limit of the union of all these algebras. Segal says that he takes the weak limit in order to include as many operators as possible that might be argued by physicists to be bounded observables. We DO NOT ADD these weak limits; this avoids the possibility that we might include in the algebra some global operators that are not norm limits of *observable* operators localised in a *bounded* region of space. This is required by the Haag–Kastler axioms, discussed in Chapter 3.

5.3 Representations with Coherent States

An extensive account of coherent states is presented by Ali, Antoine and Gazeau in Ref. 5. Consider a quantum system which has N degrees of freedom, and a Hamiltonian

$$H = \frac{1}{2}\sum_{j=1}^{N} E_j(p_j^2 + q_j^2), \tag{5.51}$$

where $E_j > 0$, $j = 1, 2, \ldots, N$. This, then, is the system of N harmonic oscillators. We may write H in terms of the creation and annihilation operators a_j, a_j^*, $j = 1, \ldots, N$, which are defined by

$$a_j = 2^{-\frac{1}{2}}(q_j + ip_j). \tag{5.52}$$

This gives

$$H = \sum_{j=1}^{N} E_j a_j^* a_j + \text{constant}. \tag{5.53}$$

There is a unique vacuum state Ψ_0, such that $a_j\Psi_0 = 0$ for all j and all choices of $\Psi_0 \in \Psi_0$. The constant in Eq. (5.53) can be chosen so that $H\Psi_0 = 0$. A *coherent state*, denoted $\Phi(z)$, of the harmonic oscillator, labelled by the parameters $z = \{z_j\} \in \mathbf{C}^N$, $j = 1, 2, \ldots, N$, has four famous properties, not necessarily equivalent or independent of each other. For a single oscillator, $N = 1$, denote the position operator by Q and the momentum operator by P. Then for one degree of freedom, the four properties of the coherent states are given by Ref. 5, page 2:

- They are states of minimal uncertainty, in that

$$\Delta Q(z)\Delta P(z) = \frac{1}{2}\hbar \tag{5.54}$$

where $\Delta Q(z)$ is the standard deviation of the position operator Q in the state $\Phi(z)$:

$$\Delta Q(z) := [(\Phi(z), Q^2\Phi(z)) - (\Phi(z), Q\Phi(z))^2]^{1/2}, \tag{5.55}$$

and $\Delta P(z)$ is similarly the standard deviation of the momentum operator P in the same state. For a general state, it can be shown that equality in Eq. (5.54) is replaced by $\geq$, so coherent states have indeed got minimal uncertainty.
- For any $z \in \mathbf{C}$, $\Phi(z)$ is an eigenstate of the annihilation operator a, with eigenvalue equal to z:

$$a\Phi(z) = z\Phi(z). \tag{5.56}$$

- The Heisenberg group, G, with one degree of freedom is the group generated by the unitary operators

$$U(\alpha) = \exp i\alpha Q, \quad V(\beta) = \exp i\beta P. \tag{5.57}$$

Every representation contains at least one vacuum state, Ψ_0 say. The action of $U(\alpha)$ and $V(\beta)$ on Ψ_0 generates a subrepresentation of G which possesses Ψ_0 as a cyclic vector. The coherent states of any cyclic representation of the Heisenberg group are created from the vacuum by the unitary action of the Heisenberg group, generated by Q, P and the identity I, or equally, by the creation operator a^* and annihilation operator a:

$$\Phi(z) = \exp\{za^* - \bar{z}a\}\Psi_0 := W(z)\Psi_0. \tag{5.58}$$

- There is a resolution of the identity: put $z = x + iy$.

$$I = \frac{1}{\pi} \int_{\mathbf{C}} dx dy \, I_z, \tag{5.59}$$

where I_z is the projection operator onto the state $\Phi(z)$. The coherent states thus constitute an overcomplete family.

We shall use the third of these properties, $W(z)\Psi_0$, for our choice of coherent state when there are infinitely many canonical variables in the theory, such as the free transverse electromagnetic quantum field theory with infinitely many modes.

5.4   The Segal–Bargmann Transform

The Segal–Bargmann transform, also called the coherent-state transform, is an integral transform, A say, which maps $L^2(\mathbf{R}^n)$ into the space $\mathcal{H}(\mathbf{C}^n)$ of holomorphic functions on $\mathbf{C}^n$: for $f \in L^2(\mathbf{R}^n)$ and $z \in \mathbf{C}^n$; then

$$A(f) = \int_{\mathbf{R}^n} A(z, x) f(x) d^n x, \tag{5.60}$$

where

$$A(z, x) = (2\pi)^{-\frac{n}{4}} \exp\left[-\frac{1}{4} \sum_{j=1}^{n} (2z_j^2 + x_j^2) + \sum_{j=1}^{n} x_j z_j \right]. \tag{5.61}$$

The standard heat-kernel on $\boldsymbol{R}^n$,

$$\rho_t(x) = (2\pi t)^{-n/2} \exp\left[-\sum_{j=1}^{n} x_j^2/2t\right] \qquad (5.62)$$

has an analytic continuation $\rho_t(z)$ to $\boldsymbol{C}^n$, which is related to $A(z, x)$:

$$A(z, x) = (\rho_t(x))^{-1/2} \rho_t(z - x). \qquad (5.63)$$

Bargmann [15] proved that the map A is an isometric isomorphism of $L^2(\boldsymbol{R}^n)$ onto the Hilbert space $\mathcal{H}(\boldsymbol{C}^n) \cap L^2(\boldsymbol{C}^n, \mu)$, where μ is the Gauss measure

$$d\mu(z) = \pi^{-n} \exp\left[-\sum_{j=1}^{n} |z_j|^2\right] dx\, dy. \qquad (5.64)$$

Segal was once heard saying that Bargmann was able to construct the proof of the finite-dimensional case from his, Segal's, earlier proof [54, 55] in the infinite-dimensional case.

Let us now describe a representation that might be a good model of an *interacting* particle.

Our reference for some of this material is the paper of Araki [7]. We start with an automorphism of the algebra of observables given by a Shale transformation R of the form given in Eq. (5.45) in which we take T to be the identity. Let us denote the representation we get by W_Φ. The states of the representation are called *coherent states*, and W_Φ is equivalent to the Fock representation if and only if $\Phi \in \mathcal{K}_1$.

We shall follow Doplicher and Roberts and generalise this construction, by using an *endomorphism* of the algebra instead of an automorphism, such as that used by Shale. An invertible endomorphism is an automorphism, so this is a true generalisation: it leads to a possibly reducible representation of the algebra, which is shown in the next chapter. Instead of adding a field Φ which is complex-valued, we add a vector field to the connection ∂_μ which lies in the algebra of $n \times n$ matrices. The states we get show that a symmetry group $SU(n)$ together with superselection rules, has then been introduced.

We now construct new representations by choosing T to act at time zero on the free electromagnetic field F. T is the identity operator acting on fields at $\boldsymbol{x}$ inside a ball of radius $a > 0$, so that $\|\boldsymbol{x}\| \le a$ in $\boldsymbol{R}^3$; outside this ball, we choose T to be a space-dependent rotation, going smoothly from the identity on the surface of the sphere $\|\boldsymbol{x}\| = a$ to a complete 2π

rotation at the radius $\|x\| = b > a$, and T remains the identity on fields located at x with $\|x\| > b$. At points in between the inner sphere and the outer one, we rotate the fields $\boldsymbol{E}$, $\boldsymbol{H}$ about 0_x, the line joining x to the origin. The states might be called "Higgs" states.

We conjecture that for this choice of T, the operator $1 - |T|$ is not of Hilbert–Schmidt class, and that we therefore get a non-Fock representation of the free electromagnetic field. In this representation, a quantum of the electromagnetic field of high energy will have a short wave length, and if it is less than a it could reach inside the Higgs particle, where it feels it is in a vacuum. We thus conjecture that the interaction between such a Higgs particle and a photon will go to zero at high energy.

Chapter 6

Euclidean Electrodynamics

This chapter owes a lot to the presentation given by Simon [58]. The free fields E, H obey the Wightman axioms, and so the energy operator in the theory is bounded below. Moreover, the vacuum expectation values at time $t = 0$ have an analytic continuation in t in the upper-half-plane, $\operatorname{Im} t > 0$. The continuation from $(0, x)$ to (it, x) yields a Euclidean tensor field [76] which describes a non-quantum, that is, a classical field. We call it a Nelson field because it was he who showed us how to do the case for the scalar field [52]. In this chapter, we shall first describe the massive relativistic free quantised scalar field. In this case, the Wightman functions of the field define, by analytic continuation to imaginary time, a Euclidean real field obeying Nelson's Markov property. We then show that a Hermitian, massless free Wightman field of spin one, namely the free electromagnetic field, defines a vector Euclidean field obeying a similar Markov property. In these cases, the Nelson field uniquely defines the corresponding Wightman field, up to unitary equivalence.

6.1   Some Probability Theory

If A has a largest eigenvalue which is simple, then we can find it, and also can construct the eigenvector, by the method of matrix squaring:

$$A^{2^m} \psi \sim c\lambda^{2^m} \psi_0 \tag{6.1}$$

as m tends to infinity, where $c = \langle \psi_0, \psi \rangle$. This is a special case of Markov's theorem: we are interested in finding the possible stable distributions of frequencies, that is, a vector ψ_0 such that $A\psi_0 = \psi_0$. In this case, 1 is the largest eigenvalue, and

$$\psi_0 = \lim_{m \to \infty} A^m \psi \qquad (6.2)$$

for any ψ, if all A_{ij} are positive.

An $n \times n$ matrix P with $P_{ij} \geq 0$ and $\sum_j P_{ij} = 1$ is called a *Markov matrix*. Here, P_{ij} is the probability that the state will be in the state i at time $t + 1$, given that it is in the state j at time t. Any such matrix defines a *stationary Markov process with discrete time*. To see what this means, we now develop the theory of conditioning. We start in discrete time, with the case where the space of outcomes is finite.

Let Ω be a space of n elements, $n < \infty$, and denote by $\mathcal{B}$ the set of subsets of Ω; this is the set of *events*. Suppose that P is a non-negative measure, such that $P(\Omega) = 1$, defined on $\mathcal{B}$. An event B can happen if $P(B)$ is positive. If so, we can define the conditional probability of the event A, given B, to be

Definition 6.1 $P(A \mid B) := P(A \cap B)/P(B)$.

It is clear that $P(A \mid B)$ is a probability measure on Ω: indeed, for any set A, $P(A \mid B) \geq 0$, and

$$\sum_{\omega \in \Omega} P(\omega \mid B) = \sum_{\omega \in B} P(\omega \mid B) + \sum_{\omega \notin B} P(\omega \mid B) \qquad (6.3)$$

Thus,

$$\sum_{\omega \in \Omega} P(\omega \mid B) = \sum_{\omega \in B} P(\omega \mid B)$$

$$= \sum_{\omega \in B} P(\omega)/P(B)$$

$$= 1.$$

Now let f be a random variable; that is, f is a measurable function on $(\Omega, \mathcal{B}, P)$. Then the *conditional expectation* of f, given B, is

$$E(f \mid B) = \sum_\omega P(\omega \mid B) f(\omega). \qquad (6.4)$$

We can also find the conditional probability, given that B did *not* happen, $P(\omega \cap B^c)/P(B^c)$, where $B^c = \{\omega \notin B\}$; the corresponding conditional expectation is then

$$E(f \mid B^c) = \sum_\omega P(\omega \mid B^c) f(\omega) \qquad (6.5)$$

We may regard $(E(f \mid B), E(F \mid B^c))$ as defining a simple function on Ω, equal to $E(f \mid B)$ if $\omega \in B$ and to $E(f \mid B^c)$ if $\omega \notin B$. In this form, the concept may be generalised: let $B_1, \ldots, B_N$, $N \leq \infty$, be disjoint measurable sets such that $P(B_j) \neq 0, j = 1, \ldots, N$, and $P(\cup_j B_j) = 1$. These sets generate a Boolean ring (with 2^N elements), say $\mathcal{B}_0$. For each random variable f, the functions on Ω defined by

$$F_f(\omega) = E(f \mid B_i) \tag{6.6}$$

if $\omega \in B_i$, are measurable with respect to $\mathcal{B}_0$. They take constant values $E(f \mid B_i)$ on each B_i, and so can be written

$$F_f(\omega) = \sum_i \chi_{B_i}(\omega) c_i \tag{6.7}$$

where $c_i = E(f \mid B_i)$. In fact, every function F, measurable relative to $\mathcal{B}_0$, has the form of F_f given by Eq. (6.7) for some choice of f and c_i, $i = 1, \ldots, N$. This space is denoted by L^0.

The theory described so far is also valid more generally; we get similar results when Ω is a space which is not necessarily countable, but carries a measure μ with $\mu(\Omega) = 1$. The assumptions made so far are special cases of this. The results above can be generalised: we simply replace $\sum_\omega P$ by $\int_\Omega d\mu$, $P(B_j)$ by $\mu(B_j)$, $P(\omega \mid B)$ by $\mu(\omega \mid B)$, and so on. We use this more general theory in the following.

Thus, suppose that $f \in L^2(\Omega, \mathcal{B}, \mu)$; then the map

$$f \to F_f = \{E(f \mid B_i)\}, \quad i = 1, \ldots, N \tag{6.8}$$

is a map to $L^2(\Omega, \mathcal{B}_0, \mu)$ which leaves each element of $L^2(\Omega, \mathcal{B}_0, \mu)$ unchanged. Indeed, suppose that f has the form

$$f(\omega) = \sum_i c_i \chi_{B_i}(\omega). \tag{6.9}$$

Then

$$E(f \mid B_i) = \int f(\omega) d\mu(\omega \mid B_i)$$

$$= c_i \int_{\omega \in B_i} d\mu(\omega \mid B_i)$$

$$= c_i$$

This shows that $F_f(\omega) = f(\omega)$ up to sets of μ-measure zero if $F \in L^0(\Omega, \mathcal{B}_0, \mu)$. Moreover, if $g \in L^2(\Omega, \mathcal{B}, \mu)$, then

$$\langle g, F_f \rangle = \int g(\omega) d\mu(g) \sum_i \chi_{B_i}(\omega) \sum_{\omega'} d\mu(\omega' \mid B_i) f(\omega')$$

$$= \sum_i \int_{\omega \in B_i} g(\omega) d\mu \int_{\omega' \in B_i} d\mu(\omega') f(\omega') / \mu(B_i)$$

$$= \langle F_g, f \rangle$$

by symmetry. Hence the map $F : f \mapsto F_f$ obeys $F^2 = F$ and $F = F^*$; that is, F is the orthogonal projection onto $L^2(\Omega, \mathcal{B}_0, \mu)$. This motivates the following definition [58]. Let $(\Omega, \mathcal{B}, \mu)$ be a probability space, and $f \in L^2(\Omega, \mathcal{B}, \mu)$. Let $\mathcal{B}_0 \subseteq \mathcal{B}$ be a sub-σ-ring. Then the projection of f onto the subspace $L^2(\Omega, \mathcal{B}_0, \mu)$ is called the *conditional expectation* of the function f, *given* $\mathcal{B}_0$. The conditional expectation averages over the variables not dependent on the variables in $\mathcal{B}_0$.

Let W_0 be the Fock representation, and let Φ be a transverse solution to the Maxwell equations. Let W_Φ be the Weyl operators obtained by using the fields given Eqs. (5.47)–(5.50). We get the *coherent states* [43], and by Theorem 6 the representation is equivalent to the Fock representation if and only if $\Phi \in \mathcal{K}_1$. When Φ does not lie in $\mathcal{K}_1$, we still call the representation a coherent-state representation, but it is not equivalent to W_0.

The relation of the Haag–Kastler axioms to those of Wightman is not clear for a general Wightman theory, but for any free boson field a key result due to Slawny [59] suggests a natural way to construct a set of local C*-algebras which obey the Haag–Kastler axioms.

We now define the Haag–Kastler field [8, 36, 37] for a system of Segal operators. Let $\mathcal{O}$ be a bounded open set in $\mathbf{R}^4$, of the form of the intersection of the interiors of a forward and a backward light cone. The surfaces of the cones themselves intersect in a two-dimensional set, which spans a three-dimensional region D inside both cones. Let f and g be continuous functions on D. Then the local C*-algebra $\mathcal{A}(\mathcal{O})$ is the completion, in the Slawny norm, of the Segal *-algebra generated by such f and g. The algebra of all observables, $\mathcal{A}$, is then the completion of the inductive limit of the union over $\mathcal{O}$ of all the local algebras; in this union, we of course identify the elements of $\mathcal{A}(\mathcal{O}_1)$ with a subalgebra of any $\mathcal{A}(\mathcal{O})$ whenever $\mathcal{O}_1 \subset \mathcal{O}$. That is, we form the union of all the algebras $\mathcal{A}(\mathcal{O})$, and then complete it in the Slawny norm. The algebra assigned to an arbitrary connected open region in $\mathbf{R}^4$ is the completion of the union of all $\mathcal{A}(\mathcal{O})$, $\mathcal{O}$ being a subset of the region. Note that this is not exactly the same set of operators as defined and used by Segal; in our algebra, every element is either localised in a bounded set $\mathcal{O}$, or, if it is obtained from the inductive limit, it is the norm limit of a sequence of localised elements. Segal adds the weak limits of some local elements, as well as including all square integrable solutions in his algebra.

The C^*-algebra $\mathcal{A}$ defined by the previous paragraph obeys the axioms of Haag and Kastler [37] except that they assumed that the Poincaré group acted on $\mathcal{A}$ norm-continuously, which we do not. The free field satisfies one more axiom, the split property of Doplicher and Roberts [30].

6.2 Markov Processes

Suppose that a large community is served by a number, n, of competitive shops. At a given weekend, t, where $t \in \mathbf{Z}$, the state of the system is given by a vector $\psi = (\psi_1, \ldots, \psi_n)$, where ψ_j is the proportion of shoppers using shop j, $j = 1, \ldots, n$. Clearly, $\psi_j \geq 0$ for all j, and $\sum_j \psi_j = 1$. Suppose that buying habits are not hard-and-fast, but have a stochastic element. Suppose that $p_{i,j}, 1 \leq i, j \leq n$, denotes the probability that a shopper will buy from the shop i, given that in the previous week he shopped in shop j. The Markov property is the assertion that the probability that the shopper chooses i, given that in the previous week he chose shop j, is independent of where he shopped before the last week.

The matrix with components $p_{i,j}$ is called the transition matrix. We obviously have

$$p_{i,j} \geq 0 \qquad (6.10)$$

for all $0 \leq, i, j \leq n$, and

$$\sum_i p_{i,j} = 1 \qquad (6.11)$$

for all j. A state ψ is transformed to $P\psi$ a week later. A state ψ_0 such that $P\psi_0 = \psi_0$ is called stationary. It is obvious that a matrix, say A, has non-negative elements if and only if it is *positivity preserving*: that is $\psi_j \geq 0$ for all j implies that $(A\psi)_j \geq 0$. The stronger condition, that a matrix has *positive* elements, leads to a useful result.

Theorem 7 (The Perron–Frobenius Theorem) *Let A be a symmetric matrix with positive elements. Then the largest eigenvalue of A is simple, and the corresponding eigenvector may be chosen with non-negative components.*

Proof Let λ be the largest eigenvalue. Then

$$\lambda = \sup\langle\psi, A\psi\rangle/\langle\psi, \psi\rangle \tag{6.12}$$

$$\geq \sum_{i,j} A_{ij}/n \tag{6.13}$$

by choosing $\psi = (1, 1, \ldots, 1)$. Hence the largest eigenvalue is positive. Now let ψ be any eigenstate with eigenvalue λ, and denote by $|\psi_0|$ the vector with components $|\psi_j|$. Then

$$\langle|\psi_0|, A|\psi_0|\rangle = \sum_{i,j} A_{ij}|\psi_i|\,|\psi_j|$$

$$\geq \left|\sum A_{ij}\bar{\psi}_i\psi_j\right|$$

$$= \lambda\langle\psi, \psi\rangle$$

$$= \lambda\|\,|\psi_0|\,\|^2.$$

Hence we have equality,

$$\langle|\psi_0|, A|\psi_0|\rangle = \lambda\|\,|\psi_0|\,\|^2 \tag{6.14}$$

by the supremum property, Eq. (6.12), and we have also proved that $|\psi_0|$ is an eigenvector of A. No component of ψ can be zero:

$$\lambda|\psi_i| = \sum_j A_{ij}|\psi_j| > 0, \tag{6.15}$$

since not all components ψ_j are zero. Finally, the inequality

$$|A_{ij}\bar{\psi}_i\psi_j| \leq A_{ij}|\psi_i|\,|\psi_j| \tag{6.16}$$

for all i, j must be an equality, since if for one pair, i, j, we had a true inequality, then summing over i and j could not reach equality for $\sum_{ij} A_{ij}|\psi_i|\,|\psi_j|$ and $|\sum_{ij} A_{ij}\bar{\psi}_i\psi_j|$. Since $A_{ij} \neq 0$ we may cancel this factor in Eq. (6.16), to get

$$\bar{\psi}_i\psi_j = |\psi_i|\,|\psi_j|. \tag{6.17}$$

It follows from this that the complex phases of all the components, $\psi_1, \ldots, \psi_n$, are equal, and so the vector ψ is a complex multiple of $|\psi_0|$. Hence the largest eigenvalue of A is simple. This proves the theorem.

We shall see that $\exp\{-Ht\}$, where H is the Hamiltonian of a free quantum field, is positivity preserving, leading to the uniqueness of the ground state of H, by the generalisation of the Perron–Frobenius theorem to infinite dimensions. We have seen that in $\boldsymbol{R}^n$ the proof actually uses the fact that our operator, A in the above, is actually positivity *improving*: $\psi_j \geq 0$ implies that $(A\psi)_j > 0$.

6.3 Independence

Let μ be a probability measure on a space Ω. *Events* are represented by measurable subsets of Ω. We say that two events, $A \subseteq \Omega$ and $B \subseteq \Omega$ are *independent* if $\mu(A \cap B) = \mu(A)\mu(B)$. We say that two random variables $f, g \in L^0(\Omega)$ are *independent* if the set $\{\omega \in \Omega : a \leq f(\omega) \leq b\}$ is independent of the set $\{\omega \in \Omega : c \leq g(\omega) \leq d\}$, for all real values of a, b, c, d. This can be shown to be equivalent to saying that the smallest σ-ring, call it $\mathcal{B}(f)$, relative to which f is measurable, is independent of (every set in) the smallest σ-ring $\mathcal{B}(g)$ relative to which g is measurable.

Similarly, given $\mathcal{B}_1$ and $\mathcal{B}_2$, both subsets of $\mathcal{B}$, we say that $\mathcal{B}_1$ and $\mathcal{B}_2$ are independent if any finite collection of sets in $\mathcal{B}_1$ is independent of any finite collection of sets in $\mathcal{B}_2$. This is equivalent to saying that $L^0(\mathcal{B}_1)$ is independent of $L^0(\mathcal{B}_2)$.

If $f \in L^0(\mathcal{B})$ is independent of an event B, then

$$E(f \mid B) = \sum_\omega P(\omega \mid B) f(\omega)$$

$$= \sum_j f_j P(A_j \mid B) \quad \text{where } A_j = \{\omega : f(\omega) = f_j\}$$

$$= \sum_j f_j P(A_j \cap B)/P(B)$$

$$= \sum_j f_j P(A_j) P(B)/P(B)$$

$$= E(f).$$

Hence, if $\mathcal{B}_0$ is the Borel field generated by $B_1, \ldots, B_n$, we have

$$F_f(\omega \mid \mathcal{B}_0) = \sum_i \chi(B_i) E(f \mid B_i)$$

$$= E(f) \sum_i \chi(B_i)$$

$$= E(f)$$

We therefore see that the following useful criterion holds: two σ-rings $\mathcal{B}_1$ and $\mathcal{B}_2$ are independent if and only if

$$E(f_1 \mid \mathcal{B}_2) = E(f_1) \quad \text{for all } f_1 \in L^0(\mathcal{B}_1) \tag{6.18}$$

and

$$E(f_2 \mid \mathcal{B}_1) = E(f_2) \quad \text{for all } f_2 \in L^0(\mathcal{B}_2). \tag{6.19}$$

6.4   Markov Processes with Stationary Distributions

Let n be a positive integer. An $n \times n$ matrix, A say, is called a *transition matrix* if

$$A_{ij} \geq 0 \quad \text{for all } i \text{ and } j, \tag{6.20}$$

$$\sum_j A_{ij} = 1 \quad \text{for all } i. \tag{6.21}$$

If $\psi_i \geq 0$ and $\sum_i \psi_i = 1$, (ψ is a probability) then so is $A\psi$, and also are $A^t\psi$ for all integers t. Suppose that the (classical) system S can be found in, at most, n possible states; we can then label the state of the system by the integer $i \in S := \{1 \leq i \leq n\}$. Suppose we are told that, if at a time t, an integer, the system is in state i, then at time $t+1$ the probability that it has moved to state j is A_{ij}. Suppose that at time t, we are told that the probability that the state is i, is ψ_i:

$$\text{Prob}\{X = i\} = \psi_i. \tag{6.22}$$

Then the probability that the system is in state j at time $t+1$ is $\sum_i A_{ij}\psi_i$. We need to find one sample space, with a measure on it, so that all the variables $X(t)$ are random variables on the same space: we must construct a measure space and a stochastic process $\{X_t\}$ with values (at each integer time $t \geq 0$) in the set S, such that

$$\text{Prob}\{X_t = i\} = (A^t\psi)_i, \quad t \in \{0, 1, 2, \ldots\} \tag{6.23}$$

We have adopted the conventions of taking the starting point to be time zero, the time interval to be unity, and the distribution at time zero to be given by ψ. The sample space is the set of sequences $\{X_t : t \geq 0\}$ of integers j from the set $\{1 \leq j \leq n\}$. We write $\mathcal{B}_t$ for the σ-ring generated by X_t, and $\mathcal{B}_{\leq t}$ for the smallest σ-ring for which $\{X_s : s \leq t\}$ are measurable. In the case here, these σ-rings are generated by the collection of subsets of the form $\{C_0, C_1, \ldots, C_r, S, S, S, \ldots\}$: here, $C_0, \ldots, C_r$ are $r+1$ subsets of S, and the rest of the sequence consists of copies of the whole space S.

The process is a Markov process, in that the distribution of X_{t+1} depends only on the "state" at time t; in terms of the conditional expectations, we have

$$E(f(X_{t+1} \mid \mathcal{B}_{\leq t})) = E(f(X_{t+1} \mid \mathcal{B}_t)) \tag{6.24}$$

for every function f.

We omit the theory of Wiener measure, and go on to the cases of the free relativistic quantised fields of mass zero and spin zero and spin one. First, a bit of general theory.

6.5   Linear Processes

Let $(\Omega, \mathcal{B}, \mu)$ be a probability space; let $f : \Omega \to \mathbf{R}$ be a random variable. We denote by $L^0(\Omega, \mathcal{B})$ the set of $\mathcal{B}$-measurable functions on Ω. If $f \in L^k$ $(\Omega, \mathcal{B}, \mu)$ for all positive integers k, we say that all the moments of f exist: $m_n(f) = E(f^n)$. The characteristic function

$$E(\exp(i\lambda f)) = G_f(\lambda) \tag{6.25}$$

exists for any random variable f. It has the properties

(1) $G_f(0) = 1$.
(2) $G_f(\lambda)$ is continuous.
(3) $G_f(\lambda)$ is of positive type:

$$\sum_{ij} \bar{\alpha}_i \alpha_j G_f(\lambda_i - \lambda_j) \geq 0$$

for all real $\lambda_1, \ldots, \lambda_n$ and all complex $\alpha_1, \ldots, \alpha_n$.

Conversely, Bochner's theorem asserts that, given any function $G(\lambda)$ satisfying (1), (2) and (3), then there exists a probability measure μ on $\mathbf{R}$ such that $G(\lambda) = \int \exp(i\lambda x) d\mu(x)$. If G is the characteristic function of a random variable f on $(\Omega, \mathcal{B}, \mu)$, then the corresponding measure, μ, on $\mathbf{R}$, is called the *probability distribution* of f, and is written as $\mu_f(x)$.

If $G_f(\lambda)$ is an analytic function of λ in an interval $\{\lambda : |\lambda| < \epsilon\}$, then the moments $m_n(f)$ are given by

$$G_f(\lambda) = \sum_{n=0}^{\infty} \frac{(i\lambda)^n}{n!} m_n(f). \tag{6.26}$$

The *cumulants* $\kappa_n(f)$ of f are defined inductively in terms of the moments by

$$m_1 = \kappa_1 \tag{6.27}$$

$$m_n = \kappa_n + \sum_{I=(i_1,\dots,i_k)} \kappa_{i_1} \times \cdots \times \kappa_{i_k} \tag{6.28}$$

where the sum is over all partitions of $\{1, 2, \dots, n\}$ into k parts, where $1 < k \leq n$ and $i_s, 1 \leq s \leq k$, is the number of elements in the sth part. The cumulants are related to the characteristic function G_f by

$$\log G_f(-i\lambda) = \sum_{n=1}^{\infty} \frac{\lambda^n}{n!} \kappa_n. \tag{6.29}$$

Clearly, the probability distribution of f, μ_f, has the property

$$\text{Prob}\{-\infty < f \leq x\} = \mu_f^{-1}(-\infty, x]$$

$$= \int_{-\infty}^{x} d\mu_f$$

$$= \mu_f(-\infty, x].$$

We say that a random variable f has a Gaussian distribution if there are numbers $m \in \mathbf{R}$ and $\sigma > 0$ such that

$$\text{Prob}\{-\infty < f \leq x\} = \frac{1}{\sigma(2\pi)^{1/2}} \int_{-\infty}^{x} \exp \frac{-(y-m)^2}{2\sigma^2} dy. \tag{6.30}$$

Then the mean of f is m and the variance is σ^2. One shows that, if f is Gaussian, then

$$G_f(\lambda) = \exp\{i\lambda m\} \exp\{-\sigma^2 \lambda^2/2\}, \tag{6.31}$$

so that, in this case, $\kappa_n = 0$ if $n > 2$; that is, the cumulants of order more than 2 vanish. Conversely, given a mean m and positive standard deviation σ, there is a unique random variable f such that its first mean is m and its variance is σ^2, AND its cumulants of order greater than 2 vanish. Uniqueness here means that any other random variable, on any other space $\{\Omega', \mathcal{B}', \mu'\}$, and with the same moments, has the same probability distribution as f.

A finite set, $\{f_j\}$, of random variables is said to be *jointly Gaussian*, if any of the following equivalent statements is true:

(1) Every finite real linear sum $\sum_i \alpha_i f_i$ is Gaussian.
(2) The joint characteristic function has the form

$$E\left\{\exp\left(i\sum \lambda_i f_i\right)\right\} = \exp\left(i\sum \lambda_j \mu_j\right) \exp\left\{-\frac{1}{2}\sum_{i,j} A_{ij}\lambda_i\lambda_j\right\}.$$

(6.32)

(3) The joint cumulants $\kappa_n(i_1,\ldots,i_n)$ vanish for $n \geq 3$. They are defined inductively by

$$E(f_j) = \kappa_1(j)$$

$$E(f_{i_1},\ldots,f_{i_n}) = \kappa_n(i_1,\ldots,i_n) + \sum_I \kappa(j_1,\ldots)\ldots\kappa(\ldots,j_n)$$

where I is a general partition into parts of disjoint parts of $1,2,\ldots,n$.

An arbitrary collection $\{f_i\}_{i\in I}$ of random variables is said to be Gaussian if every finite subset is jointly Gaussian.

6.6 Random Distributions

We now try to specialize the above discussion to random distributions, in the Gaussian case. Let V be a vector space over $\mathbf{R}$. A linear stochastic process over V consists of a probability space $(\Omega,\mathcal{B},\mu)$ and a linear map $\Phi : V \to L^0(\Omega,\mathcal{B})$. Thus, Φ satisfies

$$\Phi(\lambda_1 v_1 + \lambda_2 v_2) = \lambda_1\Phi(v_1) + \lambda_2\Phi(v_2).$$

(6.33)

When V is a topological vector space, one often imposes a continuity condition on Φ, in the form: if $v_\alpha \to v$ in V as $\alpha \to \infty$, then $\Phi(v_\alpha) \to \Phi(v)$ in measure as $\alpha \to \infty$. We say that $\Phi_\alpha \to \Phi$ in measure if for every $\epsilon > 0$,

$$\mu\{\omega : |\Phi_\alpha(\omega) - \Phi(\omega)| > \epsilon\} \to 0.$$

(6.34)

We shall be interested in *random distributions*, that is, a continuous linear process Φ over $\mathcal{D}(\mathbf{R}^d)$ or $\mathcal{S}(\mathbf{R}^d)$. Here, d is the dimension of space-time. If $d = 1$ we get a stochastic process in the usual sense if for each time t there is a random variable Φ_t such that

$$\Phi(v) = \int v(t)\Phi_t dt$$

(6.35)

for each $v \in \mathcal{D}(\mathbf{R})$. This limits the random distribution to be essentially a random function.

Suppose now that Φ is a linear Gaussian process over $\mathcal{D}(\mathbf{R}^d)$ or $\mathcal{S}(\mathbf{R}^d)$. Then the joint probability distributions of all the $\Phi(h), h \in \mathcal{D}(\mathbf{R}^d)$ say, are uniquely determined by the first two joint moments, $E(\Phi(h))$ and $E(\Phi(h_1)\Phi(h_2))$. In this sense, a Gaussian process is determined by its first two moments. This is analogous to the fact that a generalised free Wightman field is determined by its one- and two-point functions.

The covariance matrix of a linear process over V defines a bilinear form on V:

$$\langle h_1, h_2 \rangle = E(\Phi(h_1)\Phi(h_2)) - E(\Phi(h_1))E(\Phi(h_2)) \qquad (6.36)$$

and this form is positive semi-definite. There is an important converse to this, namely

Theorem 8 *Let V be a real vector space and let $\langle \, , \, \rangle$ be a positive definite bilinear form on V. Then there exists a Gaussian process Φ on a probability space $(\Omega, \mathcal{B}, \mu)$ such that $\kappa_2(\Phi(v)\Phi(u)) = \langle v, u \rangle$ for all u and v in V.*

Proof For each finite collection $v_1, \ldots, v_n$, let

$$d\mu_{v_1,\ldots,v_n} = (\det\langle v_i, v_j \rangle)^{-1/2} \exp\left(-\frac{1}{2}\{b_{ij}x_i x_j\}\right) d^n x, \qquad (6.37)$$

where b is the inverse matrix to $a_{ij} = \langle v_i, v_j \rangle$. For the proof, one checks that $d\mu$ is the joint probability distribution of n Gaussian random variables of mean 0. The symmetry and compatibility conditions of Kolmogorov follow by normalising, and diagonalising these variables, which gives us n independent Gaussian random variables. These are clearly compatible as we increase n. Then, by using Kolmogorov's theorem, we may take Ω to be $\mathbf{R}^V$, and μ to be the measure on the σ-ring $\mathcal{B}$ generated by the cylinder sets given by Kolmogorov's construction.

The space $\mathbf{R}^V$ is large, and it is interesting that a much smaller space often carries measure 1. One such case follows from Minlos's theorem, which we give next. It leads instead to the space of tempered distributions, $\mathcal{S}'$, as having probability 1:

Theorem 9 *Let G be a function on $\mathcal{S}(\mathbf{R}^d)$. A necessary and sufficient condition for there to exist a cylinder-set measure, $d\mu$ on $\mathcal{S}'(\mathbf{R}^d)$, such that*

$$G(h) = \int_{\mathcal{S}'} \exp(i\Phi(h))d\mu(\Phi), \qquad (6.38)$$

for all $h \in \mathcal{S}(\mathbf{R}^d)$ is that

(1) $G(0) = 1$.
(2) *the map $h \to G(h)$ is continuous.*
(3) *For any $h_1, \ldots, h_n$ in $\mathcal{S}$ and $\alpha_1, \ldots, \alpha_n$, the following inequality holds:*

$$\sum_{ij} \bar{\alpha}_i \alpha_j G(h_i - h_j) \geq 0.$$

We do not prove Minlos's theorem.

The theorem can be extended with conditions that ensure that the measure obtained can be extended from the cylinder sets to the Borel σ-ring of $\mathcal{S}'$.

A Gaussian measure is one for which the characteristic function has the form

$$G(\varphi) = \exp\left\{-\frac{1}{2}\langle \varphi, \varphi \rangle\right\}. \tag{6.39}$$

We can apply Minlos's theorem to this case provided that the bilinear form, $\varphi \mapsto \langle \varphi, \varphi \rangle$ is continuous for $\varphi \in \mathcal{S}$: then G is continuous and so, from Minlos, the measure is concentrated on $\mathcal{S}'$.

6.7 Fock Space and its Real-Wave Realisation

This is the infinite-dimensional analogue of the known fact $\mathcal{F}(\mathbf{C}^n)$, the Fock space of n particles, is isomorphic to

$$L^2\left(\mathbf{R}^n, \exp\left\{-\frac{1}{2}\|x\|^2\right\} d^n x\right), \tag{6.40}$$

in such a way that the no-particle state Ψ_0 in $\mathcal{F}(\mathbf{C}^n)$ is the ray containing the function $(2\pi)^{-n/4} \exp(-\frac{1}{4}\|x\|^2)$.

Let $\mathcal{H}$ be a real Hilbert space and let $(\Omega, \mathcal{B}, \mu)$ be the Gaussian probability space defined by $\mathcal{H}$. Let $\mathcal{H}_c$ be the complexification of $\mathcal{H}$; for example, $\mathcal{H}_c = \mathcal{H} \oplus \mathcal{H}$ furnished with the complex structure $J(\phi \oplus \psi) = -\psi \oplus \phi$. Then $\mathcal{F}(\mathcal{H}_c)$ is isomorphic to $L^2(\Omega, \mathcal{B}, \mu)$ is such a way that Ψ_0 is mapped to 1, and the field operator $A(\phi)$ on $\mathcal{F}(\mathcal{H}_c)$ is mapped to $\Phi(\phi)$. For the proof of this, we note that $\langle \Psi_0, \exp\{iA(\phi)\}\Psi_0 \rangle = \exp\{-\frac{1}{2}\|\phi\|^2_{\mathcal{H}}\}$ as follows from the case of finitely many degrees of freedom. Then, since Ψ_0 is cyclic for the W*-algebra generated by $\exp(iA(\phi))$,

$\phi \in \mathcal{H}_c$, the result follows from the uniqueness theorem and the relation $E(\exp(i\Phi(\phi))) = \exp(-\frac{1}{2}\|\phi\|_{\mathcal{H}}^2)$.

We now define Wick ordering in the theory of probability. Let f be a random variable with finite moments. Then $: f^n :$, $n = 0, 1, 2, \ldots$, is defined recursively by

$$: f^0 := 1 \tag{6.41}$$

$$\frac{\partial}{\partial f} : f^n := n : f^{n-1} : \tag{6.42}$$

$$E(: f^n :) = 0. \tag{6.43}$$

From this definition we see that

$$: f := f - E(f) \tag{6.44}$$

$$: f^2 := f^2 - 2fE(f) + 2(E(f))^2 - E(f^2) \tag{6.45}$$

and so on. We see that $: f^n :$ is a polynomial of degree n. Properties of Wick polynomials can be found from the generating function

$$: \exp(\alpha f) := \sum_{n=0}^{\infty} \frac{\alpha^n : f^n :}{n!}. \tag{6.46}$$

Since

$$\frac{\partial}{\partial f} : \exp(\alpha f) := \alpha : \exp(\alpha f) : \tag{6.47}$$

and

$$E(\exp(\alpha f)) = 1 \tag{6.48}$$

we have

$$: \exp(\alpha f) := \exp(\alpha f)/E(\exp(\alpha f)). \tag{6.49}$$

Now suppose that f is Gaussian with mean zero and standard deviation $\sigma : E(f^2) = \sigma^2$. Then the means of odd powers of f are zero, and for even powers,

$$E(f^{2n}) = \frac{(2n)!\sigma^{2n}}{2^n n!}. \tag{6.50}$$

Then by direct computation,

$$E(\exp(\alpha f)) = \exp\left(\frac{1}{2}\alpha^2 E(f^2)\right). \tag{6.51}$$

Hence,

$$: \exp(\alpha f) := \exp\left(\alpha f - \frac{1}{2}\alpha^2 E(f^2)\right), \tag{6.52}$$

which generates Hermite polynomials.

Now suppose that f and g are such that all real linear sums $\alpha f + \beta g$, $\alpha, \beta \in \mathbf{R}$, are Gaussian. Then

Lemma 1 $E(: f^n :: g^m :) = \delta_{nm} n! (E(fg))^n$.

Proof : $\exp(\alpha f) :: \exp(\beta g) := \exp(\alpha f + \beta g) \exp\{-\frac{1}{2}(\alpha^2 \langle f^2 \rangle + \beta^2 \langle g^2 \rangle)\}$.
Hence

$$E(: e^{\alpha f} :: e^{\beta g} :) = \exp\left\{\frac{1}{2}E((\alpha f + \beta g)^2 - \alpha^2 f^2 - \beta^2 g^2)\right\}$$

$$= \exp\{\alpha\beta E(fg)\}. \tag{6.53}$$

The lemma follows by expanding the exponentials.

We now consider several random variables $f_1, \ldots, f_k$. The Wick ordering $: f_1^{n_1} \cdots f_k^{n_k} :$ is defined recursively in $n = n_1 + \cdots + n_k$, by

$$E(: f_1^{n_1} \cdots f_k^{n_k} :) = 0, \quad n > 0 \tag{6.54}$$

$$\frac{\partial}{\partial f_i} : f_1^{n_1} \cdots f_k^{n_k} := n_i : f_1^{n_1} \cdots f_i^{n_i - 1} \cdots f_k^{n_k} : \tag{6.55}$$

This leads us to the identification of the n-particle subspace (of the Fock space) with the subspace of random variables of the form $: \Phi(\varphi_1) \cdots \Phi(\varphi_k) :$. This gives us the specific identification of the n-particle state in $\mathcal{F}_n$ with products of Hermite in $L^2(\Omega, \mathcal{B}, \mu)$.

For a Gaussian random field ϕ, $W =: \Phi(f_1) \cdots \Phi(f_n) :$ is easily defined recursively as the component of $\Phi(f_1) \cdots \Phi(f_n) \in L^2(\Omega, \mathcal{B}, \mu)$ which is orthogonal to all polynomials of degree less than n. Then if $\langle f_i, f_j \rangle = 0$ when $i \neq j$, then W is the polynomial obtained by Gram–Schmidt orthogonalisation relative to the Gaussian measure. This is known to give the Hermite polynomials.

6.8   Second Quantisation

Let A be an operator on $\mathcal{H}$. We define the second-quantisation operator, $\Gamma(A)$ and $d\Gamma(A)$ to act on $\mathcal{F}(\mathcal{H})$ by

$$\Gamma(A) = 1 \oplus A \oplus A \otimes A \oplus \dots \qquad (6.56)$$

and

$$d\Gamma(A) = 0 \oplus 1 \oplus (1 \otimes A + A \otimes 1) \oplus \dots. \qquad (6.57)$$

One sees that Γ has the functorial properties

$$\Gamma(AB) = \Gamma(A)\Gamma(B) \qquad (6.58)$$

$$\Gamma(A)^* = \Gamma(A^*). \qquad (6.59)$$

Clearly, if A is unitary then so is $\Gamma(A)$; one can also prove that if A is bounded and self-adjoint, then $d\Gamma(A)$ is essentially self-adjoint on $\bigcup_n \mathcal{F}_n$. Also, if E is a self-adjoint projection, then so is $\Gamma(E)$, since $\Gamma(E)^2 = \Gamma(E^2) = \Gamma(E)$, and $\Gamma(E)^* = \Gamma(E^*) = \Gamma(E)$.

6.9   The Free Euclidean Field of Spin Zero

Let $\phi(f, t)$ denote the free quantised field of mass $m > 0$ at time $t \in \mathbf{R}$, smeared with a smooth function $f : \mathbf{R}^3 \to \mathbf{R}$. Let Ψ_0 be the vacuum state, and for each real time t, let the time evolution operator be $U(t)$. We consider the Wightman function $W(t, \boldsymbol{x})$, given by

$$\iint W(t, \boldsymbol{x} - \boldsymbol{y}) f_1(\boldsymbol{x}) f_2(\boldsymbol{y}) d^3x d^3y := \langle \Psi_0, \phi(f_1, t_1)\phi(f_2, t_2)\Psi_0 \rangle. \qquad (6.60)$$

Since $U(t)\Psi_0 = \Psi_0$ for all times t, the function $W(f_1, t_1; f_2, t_2)$ for fixed f_1, f_2 is a function only of the time difference, $t = t_2 - t_1$. Because in any Wightman theory the energy is bounded below by zero, the Fourier transform of $W(f_1, f_2; t)$ for fixed f_1, f_2

$$\widetilde{W}(f_1, f_2; E) = \frac{1}{2\pi} \int W(f_1, f_2; t) \exp\{-iEt\} dt \qquad (6.61)$$

is zero for $E < 0$. It follows that W is the boundary value of an analytic function of the time difference, t, in the upper-half-plane, $\{t : \operatorname{Im} t > 0\}$. Moreover, the Wightman function, restricted to the line $\operatorname{Im} t > 0$, $\operatorname{Re} t = 0$, is invariant under the real rotation group, $SO(4)$; this is the result of Lorentz invariance for real time. The real scalar product in four Euclidean dimensions will be denoted by

$$x \cdot y := \sum_{i=1}^{4} x_i y_i. \tag{6.62}$$

Following Nelson, we now show that the real function, call it T, of four real variables, x, s for $s > 0$, given by

$$T(s, x) = W(is, x) \tag{6.63}$$

obeys a Markov-like property.

The analytic continuation of the two-point Wightman function of the free field of spin zero to purely imaginary time leads to the Schwinger function of four variables, say $y = (y_1, \ldots, y_4)$,

$$T(y_1 - y_2) = \int \frac{\exp[ip \cdot (y_1 - y_2)]}{p \cdot p + m^2} d^4 p. \tag{6.64}$$

This is a positive definite operator, being the Fourier transform of a positive function. We now denote by $\langle f, g \rangle$ the real inner product in $L^2(\boldsymbol{R}^4, d^4x)$. Let $\mathcal{H}$ be the real Hilbert space completion of Schwartz space $\mathcal{S}_{\boldsymbol{R}}$ of real infinitely differentiable functions, with the inner product

$$(f, g) = \langle g, (-\Delta + m^2)^{-1} f \rangle. \tag{6.65}$$

Then we may define the Gaussian Euclidean field, $\Phi(f)$, over $\mathcal{H}$, with $\Omega = \mathcal{S}'(\boldsymbol{R}^4)$, $\mathcal{B} = $ cylinder-set σ-ring, μ is the Gaussian measure with covariance $\langle , \rangle$. Indeed, the set of T-functions obtained from the corresponding Wightman functions of the free field by analytic continuation to purely imaginary times, leads by a result similar to the Wightman reconstruction theorem, to a distribution-valued classical random field $\Phi(s, x)$. Note that $\mathcal{H}$, the Euclidean one-particle space, contains some distributions that are not functions. For example, $f(x) \otimes \delta(t - t_0)$ lies in $\mathcal{H}$ for all $f \in \mathcal{S}(\boldsymbol{R}^3)$. We shall show that Φ is Markovian in the time direction. Roughly, for positive s, the conditional random field, $\Phi(s, x)$, given its value at $s = 0$, is independent of the conditional random field $\Phi(s', x)$, for negative s', given the same data. This is equivalent to the following: let L be the set of continuous functions of x of compact support. Then, for a given positive value of s, the joint distribution of $\Phi(s, f)$, as f runs over L, given the values of $\Phi(0, f_1)$ for all $f_1 \in L$, is the same as the joint distribution of $\Phi(s, f)$, given the values of $\Phi(t, f_1)$ for all $f_1 \in L$ and all $t \leq 0$.

For any $\mathcal{K} \subseteq \mathcal{H}$, a closed subspace, let $\mathcal{B}(\mathcal{K})$ denote the σ-ring in $\mathcal{B}$ generated by $\Phi(f), f \in \mathcal{K}$.

Theorem 10 *Let H be a real Hilbert space, $\mathcal{K}$ be a closed subspace, and let E be the projection onto $\mathcal{K}$. Then the projection onto $L^2(\Omega, \mathcal{B}(\mathcal{K}), \mu)$ is $\Gamma(E)$.*

Proof Let $\{e_j\}, j = 1, 2, \ldots$, be an orthonormal basis for $\mathcal{K}$. Then any $F \in \operatorname{ran} \Gamma(E)$ is an L^2-limit of finite sums of Wick-ordered polynomials

$$: \Phi(e_1)^{n_1} \ldots \Phi(e_m)^{n_m} :, \tag{6.66}$$

and thus of polynomials in the $\{\Phi(e_j)\}$. Therefore such an F is in $L^2(\Omega, \mathcal{B}(\mathcal{K}))$, showing that the range of $\Gamma(E)$ lies in $L^2(\Omega, \mathcal{B}(\mathcal{K}))$:

$$\operatorname{ran} \Gamma(E) \subseteq L^2(\Omega, \mathcal{B}(\mathcal{K}), \mu). \tag{6.67}$$

Conversely, if $f \in \mathcal{K}$, then $: \exp i\Phi(f) : \in \operatorname{ran} \Gamma(E)$, as it is an L^2-convergent sum of polynomials in $\mathcal{K}^{\otimes n}$, $n = 0, 1, \ldots$. Hence $\operatorname{ran} \Gamma(E)$ contains functions of the form

$$\mathcal{A} := \{G(\Phi(f_1), \ldots, \Phi(f_n))\}, \tag{6.68}$$

for all $G \in \mathcal{S}(\boldsymbol{R}^m), f_i \in \mathcal{K}$, and all $m = 1, 2, \ldots$. Therefore $\operatorname{ran} \Gamma(E)$ contains the subspace spanned by $\mathcal{A}\mathbf{1}$, the set of all vectors obtained by acting with elements of $\mathcal{A}$ on the unit function $\mathbf{1}$. We use von Neumann's notation: if $\mathcal{B}$ is a set of bounded operators, we denote its commutant by $\mathcal{B}'$; this is the set of all bounded operators that commute with all $B \in \mathcal{B}$. Then the double commutant, $\mathcal{B}'' := (\mathcal{B}')'$ is always a weakly closed algebra of bounded operators; that is, $\mathcal{B}''$ is a W*-algebra. It can be proved to be the smallest W*-algebra containing $\mathcal{B}$.

Now, returning to the main subject, any projection in $\mathcal{A}''$ defines a measurable set in $\mathcal{B}(\mathcal{K})$, and the collection of such gives a σ-algebra, $\mathcal{B}_0$ say. For any $\Phi(f)$, its projections lie in $\mathcal{A}''$, and so for all $f \in \mathcal{K}$, $\Phi(f)$ is measurable with respect to $\mathcal{B}_0$. Since $\mathcal{B}(\mathcal{K})$ is minimal, $\mathcal{B}_0 = \mathcal{B}(\mathcal{K})$. Hence $\mathcal{A}'' = L^\infty$, all multiplication operators on $L^2(\Omega, \mathcal{B}(\mathcal{K}), \mu)$. Now, the closure of $\mathcal{A}\mathbf{1}$ contains $\mathcal{A}'' = L^\infty$, which is dense in $L^2(\Omega, \mathcal{B}(\mathcal{K}), \mu)$. Hence its closure, the range of $\Gamma(E)$, contains $L^2(\Omega, \mathcal{B}(\mathcal{K}), \mu)$, completing the proof of the theorem.

We now prove the pre-Markov property, following Simon [58] which makes the original work by Nelson [52] more accessible. Let Λ be an open or closed set in $\boldsymbol{R}^4$, let $\partial\Lambda$ be the boundary of Λ. We denote the interior of Λ, $\Lambda - \partial\Lambda$, by Λ°. Let $\mathcal{H}_\Lambda$ be the closure in $\mathcal{H}$ of distributions in $\mathcal{H}$ with support in Λ. Let e_Λ be the projection onto $\mathcal{H}_\Lambda$, and let $E_\Lambda = \Gamma(e_\Lambda)$. By the previous theorem, E_Λ is the conditional expectation relative to $\mathcal{B}(\mathcal{K})$.

Theorem 11 *Let Λ_1 and Λ_2 be closed subsets of $\mathbf{R}^4$ such that $\Lambda_1^o \cap \Lambda_2 = \emptyset$. Then*

(1) $e_{\Lambda_1} e_{\partial \Lambda_1} e_{\Lambda_2} = e_{\Lambda_1} e_{\Lambda_2}$.
(2) *If $f \in \mathcal{H}_{\Lambda_2}$ then $e_{\Lambda_1} f$ lies in $\mathcal{H}_{\partial \Lambda_1}$.*

Proof Since $e_{\Lambda_1} e_{\partial \Lambda_1} = e_{\partial \Lambda_1}$, (1) is equivalent to $e_{\partial \Lambda_1} e_{\Lambda_2} = e_{\Lambda_1} e_{\Lambda_2}$, which is the same as (2). To prove (2) we must show that if $f \in \mathcal{H}_{\Lambda_2}$, then $e_{\Lambda_1} f$ has support in $\partial \Lambda_1$. Since Λ_1 is closed, e_{Λ_1} has support in Λ_1, so we only need to prove that $\int (e_{\Lambda_1} f)(x) g(x) dx = 0$ as distribution, if $\operatorname{supp} g \subset \Lambda_1^o$ and $g \in C_0^\infty$. Now,

$$\langle e_{\Lambda_1} f, g \rangle_{L_2} = \langle e_{\Lambda_1} f, (-\Delta + m^2) g \rangle_{\mathcal{H}}$$
$$= \langle f, e_{\Lambda_1}(-\Delta + m^2) g \rangle_{\mathcal{H}}$$
$$= \langle f, (-\Delta + m^2) g \rangle_{\mathcal{H}}$$
$$= \langle f, g \rangle_{L^2}$$
$$= 0.$$

This immediately gives us the Markov property:

Theorem 12 *Let Λ_1 and Λ_2 be closed subsets of $\mathbf{R}^4$, with $\Lambda_1^o \cap \Lambda_2 = \emptyset$. Then*

(1) $E_{\Lambda_1} E_{\partial \Lambda_1} E_{\Lambda_2} = E_{\Lambda_1} E_{\Lambda_2}$.
(2) *If F is $\mathcal{B}(\mathcal{H}_{\Lambda_2})$-measurable, then the conditional expectation $E_{\Lambda_1}(F)$ is $\mathcal{B}(\mathcal{H}_{\partial \Lambda_1})$-measurable.*
(3) *If $F \in L^0(\Omega, \mathcal{B}(\mathcal{H}_{\Lambda_2}))$ then $E_{\Lambda_1} F = E_{\partial \Lambda_1} F$.*

Proof Second quantise (1) of Theorem 11 gives (1). Since $E_{\Lambda_1} E_{\partial \Lambda_1} = E_{\partial \Lambda_1}$ items 2 and 3 follow.

The following result is known as *the theorem on corridors*.

Theorem 13 *Suppose that subsets A and B of $\mathbf{R}^4$ are separated by a slab, C, of dimension three, all sets being closed. Then for any Markov field, with conditional expectation E_S for a four-dimensional subset S, we have*

$$E_A E_C E_B = E_A E_B. \tag{6.69}$$

Proof Let $D = A \cup C$ and H be the part of the boundary of C closest to B. Then

$$E_A E_C E_B = E_A E_C E_D E_B$$
$$= E_A E_H E_B \quad \text{by Theorem 12, (1)}$$
$$= E_A E_D E_B$$
$$= E_A E_B.$$

The same proof shows that we may replace the slab by its image under a continuous automorphism of $\mathbf{R}^4$. If B is bounded, we may also replace the slab C by a closed set surrounding B.

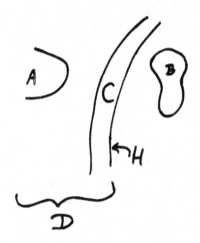

6.10   The Nelson Axioms

The free Euclidean field of mass $m > 0$, constructed by analytic continuation of the free quantised field above, obeys the following axioms, given by Nelson [52].

(1) There is a probability measure space $(\Omega, \mathcal{B}, \mu)$ and a random field Φ over

$$\mathcal{H} = L^2(\mathbf{R}^4, (p \cdot p + m^2)^{-1} d^4 p)$$

such that

- $\mathcal{B}$ is the smallest σ-field such that $\Phi(f), f \in \mathcal{H}$, is measurable.
- If $f_n \to f$ in $\mathcal{H}$, then $\Phi(f_n) \to \Phi(f)$ in measure.

(2) Covariance. There is a representation of the full Euclidean group E^4 by measure-preserving transformations T_β such that

- For each $u, v \in L^\infty(\Omega, \mathcal{B})$, the map $\beta \mapsto \int u(v \circ T_\beta) d\mu$ is measurable in β.
- $\Phi(f) \circ T_\beta = \Phi(f \circ \beta)$.

(3) The Markov property. For any closed $\Lambda \subseteq \mathbf{R}^4$ and any $\mathcal{B}(\mathcal{H}_{\mathbf{R}^4 - \Lambda})$-measurable function F, we have $E_\Lambda F = E_{\partial \Lambda} F$.

(4) Time evolution, T_t, acts ergodically on $(\Omega, \mathcal{B}, \mu)$; that is, the only sets in $\mathcal{B}(\mathrm{mod}\,\mu)$ that are invariant under T_t, are the empty set and $\mathcal{B}$.

These, together with a regularity condition, are the Nelson axioms. Nelson showed that if Φ obeys his axioms, then $E(\Phi(f_1) \cdots \Phi(f_n))$, $n = 0, 1, 2, \ldots$, are the Schwinger functions of a Wightman theory. We shall not prove this for an interacting theory, even if one exists. We shall prove the equivalence of the Nelson and Wightman axioms for any Gaussian measure. For the free quantised field, we have so far proved that the Wightman axioms imply all the Nelson axioms except ergodicity. For this, let $U(t)$ be the unitary group defined by T_t, thus:

$$(U(t)F)(\omega) := F(T_t \omega). \tag{6.70}$$

Then $U(t)$ maps the n-particle space to itself, and in fact $U(t) = \Gamma(u(t))$ where

$$(u(t)f)(x) = f(x + (t, \mathbf{0})), \quad \mathbf{0} \in \mathbf{R}^3. \tag{6.71}$$

By the Riemann–Lebesgue lemma, $u(t)$ converges weakly to zero as $t \to \infty$; hence $U(t)$ converges as $t \to \infty$ to $\Gamma(0)$, the projection onto the vacuum state. Hence $U(t)$ is ergodic.

We note that Ω can be taken as $\mathcal{S}'(\mathbf{R}^4)$ and $T_\beta(\omega)(x) = \omega(\beta^{-1}x)$, $\omega \in \mathcal{S}'$.

6.11 Reconstruction of the Hamiltonian from Nelson's Axioms

We start by assuming the Nelson axioms. So let $\mathcal{K}$ be the subspace of $\mathcal{H}$ of distributions with support on the plane $t = 0$ and let e_0 be the projection onto $\mathcal{K}$. Note that R, the reflection in the plane $t = 0$, leaves the elements of $\mathcal{K}$ pointwise invariant. Then we see that $\mathcal{K}$ consists of distributions of the form $F\delta(t)$, where F is a distribution in the three space variables.

Markov processes are naturally associated with semi-groups, as in the following theorem due to Nelson.

Theorem 14 *Let Φ be a random field obeying the Nelson axioms. Let $E_0 = \Gamma(e_0)$, and*

$$P_t = E_0 U_t E_0|_{L^2(\Omega, \mathcal{B}(\mathcal{K}), \mu)}. \tag{6.72}$$

Then $\{P_t : t \geq 0\}$ is a strongly continuous self-adjoint contraction semi-group, obeying $P_t = P_{-t}$ for all $t \geq 0$; one concludes that $P_t = \exp\{-|t|H\}$ for some self-adjoint positive operator H.

Proof Since U_t is strongly continuous, so is P_t. Since U_R leaves ran E_0 pointwise invariant, we have that $U_R E_0 = E_0 U_R = E_0$.

$$P_{-t} = E_0 U_R U_t U_R E_0$$

$$= E_0 U_t E_0$$

$$= P_t,$$

(where $E_t = \Gamma(e_t)$ and e_t is the projection onto elements of $\mathcal{H}$ with support at time t) we get from $E_t = U_t E_0 U_{-t}$ that $U_t E_s = E_{t+s} U_t$. Also, by the Markov property we have, if $t \geq 0$ and $s \geq 0$:

$$E_{-t} E_0 E_s = E_{-t} E_{(-\infty,0]} E_0 E_{[0,\infty]} E_s$$

$$= E_{-t} E_{(-\infty,0]} E_{[0,\infty]} E_s$$

$$= E_{-t} E_s.$$

We now prove that $P_t P_s = P_{t+s}$. For

$$P_t P_s = E_0^2 U_t E_0 U_s E_0$$

$$= E_0 U_t E_{-t} E_0 U_s E_0$$

$$= E_0 U_t E_{-t} E_s U_s E_0$$

$$= E_0^2 U_t E_s U_s E_0$$

$$= E_0 U_t E_s U_s E_0$$

$$= E_0 U_t U_s E_0^2$$

$$= P_{t+s}.$$

Clearly, $P_0 = 1$ and $\|P_t\| \leq 1$, so $P_t = \exp\{-\|t\|H\}$, and the theorem is proved.

Let us apply this result to the free Euclidean field of spin zero. There, $E_0 = \Gamma(e_0)$ and $U_t = \Gamma(u_t)$, where u_t is time displacement in $\mathcal{H}$. Thus $P_t = \Gamma(p_t)$ where $p_t = e_0 u_t e_0 | \mathcal{K}$. The scalar product in $\mathcal{K}$ between $F(\boldsymbol{x})\delta(t)$ and $G(\boldsymbol{x})\delta(t)$ is

$$\int \frac{\bar{\tilde{G}}(\boldsymbol{p})\tilde{F}(\boldsymbol{p})d^4 p}{p_0^2 + \boldsymbol{p}^2 + m^2} = \pi \int \frac{\bar{\tilde{G}}(\boldsymbol{p})\tilde{F}(\boldsymbol{p})d^3 p}{(\boldsymbol{p}^2 + m^2)^{1/2}}. \tag{6.73}$$

Hence $\mathcal{K}$ may be identified with the relativistic one-particle space. The formula

$$\int \frac{\exp(i\rho s)d\rho}{\rho^2 + a^2} = a^{-1}\pi e^{-a|s|} \tag{6.74}$$

implies that

$$\int \frac{\exp(itE)dE d^3 p}{E^2 + \boldsymbol{p}^2 + m^2}\bar{\tilde{G}}(\boldsymbol{p})F(\boldsymbol{p}) = \pi \int \frac{\exp(-t(\boldsymbol{p}^2 + m^2)^{1/2}\bar{\tilde{G}}F d^3 p}{(\boldsymbol{p}^2 + m^2)^{1/2}} \tag{6.75}$$

which identifies $e_0 u_t e_0$ with $\exp(-|t|h)$, where h is the one-particle Hamiltonian. Indeed, $(u_t, \mathcal{H})$ is the minimal dilation of the semigroup $\{\exp(-|t|h), \mathcal{K}\}$. Thus, $E_0 U_t E_0 = \exp(-|t|H_0)$, where H_0 is the energy of the free field. Thus, for any free massive particle, the Nelson axioms imply the Wightman axioms.

Chapter 7

Models

The construction of Wightman theories with interaction in $1 + 1$ dimensions [34] means that it has not seemed necessary to consider the idea introduced in this book for theories in two-dimensional space-time. However, in view of the difficulty, if not the impossibility, of constructing an interacting Wightman theory in four space-time dimensions, it is worthwhile pointing out the following model.

7.1 A Model in $1 + 1$ Dimensions

Consider a Wightman theory of a scalar massless free field $\hat{\phi}(t, x)$ in $1 + 1$ dimensions, with vacuum state Ψ_0. This does not exist, in that its two-point function,

$$W(s, x, t, y) := \langle \Psi_0, \hat{\phi}(s, x)\hat{\phi}(t, y)\Psi_0 \rangle \tag{7.1}$$

is not a positive semi-definite distribution. Nevertheless, the space-time derivatives, $\phi_\mu := \partial_\mu \phi, \mu = 0, 1$, do give a theory obeying the Wightman axioms in two space-time dimensions. We take these derivatives to define the observable Wightman fields. The field makes sense at sharp time, such as $s = 0, t = 0$. These time-zero fields obey a form of the canonical commutation relations which can be written in Segal form. We [64] get a Haag–Kastler field, in that it can be shown to have the Haag–Kastler properties, $(1), \ldots, (5)$. Property (6) was shown later by Buchholz and Wichmann [25] for the free massive field, and by Buchholz and Jacobi [24] for the massless case. We consider a new representation, denoted ϕ_σ, of the form

$$\partial_x \phi_\sigma = \partial_x \phi + \partial_x \varphi$$

$$\partial_t \phi_\sigma = \partial_t \phi + \varpi. \tag{7.2}$$

Here, φ and ϖ are real-valued smooth functions, and are such that $\partial_x \varphi$ and ϖ have compact support. This is a special type of Shale transformation, with $T = 1$. By Shale's theorem, the representation given by ϕ_σ is equivalent to the Fock representation if and only if the classical solution determined by the initial values φ, ϖ lies in the one-particle space. In the general case, the states of the representation given by Eq. (7.2) are said to be *coherent states* [43]. We showed in Ref. 64 that there exists a two-parameter family of superselection sectors; these can be any pair of real numbers, which we denoted by Q and Q'. If Q and Q' are both zero, then the automorphism is the identity, and so is obviously spatial, and we do not get a new representation. The set of φ considered in Ref. 64 was such that $\partial_x \varphi \in \mathcal{D}$, and the set of ϖ was $\mathcal{D}$ itself, the space of Schwartz. The representations with non-zero values of either Q or Q' are obtained by using automorphisms of the free field which are not unitarily implementable.

Consider, for example, the choice of $Q = 1$, $Q' = 1$. We now find examples of φ, ϖ giving these values. A general solution to the free wave equation can be written as the sum of a left-going wave f_L and a right-going wave f_R:

$$f(t, x) = f_L(t + x) + f_R(t - x). \qquad (7.3)$$

We choose $f_R = 0$, so we have a left-going wave, f_L; we choose a smooth $\varphi = f_L$ and $\varpi = \partial_t \varphi$, where $\varphi(x) = 1$ if x is sufficiently large, and $\varphi(x) = 0$ if x is sufficiently negative. We can think of this state (up to unitary equivalence) as given by a positive smooth function $\varphi(x)$ of compact support which rises from zero at minus infinity to one at plus infinity: it possesses a bump of size 1 in its values across its support. Since φ is not in the one-particle space, this leads to a representation, given by Eq. (7.2), not in the Fock representation. The parameters Q, Q' are both equal to 1:

$$Q = \lim_{x \to \infty} (\phi(x) - \phi(-x)) = 1$$

$$Q' = \lim_{x \to \infty} \left(\int^x \varpi(x) dx - \int^{-x} \varpi(x) dx \right) = 1. \qquad (7.4)$$

Similarly, we can construct an automorphism leading to a right-going wave of charges $Q = Q' = -1$; this commutes with the automorphism leading to charges $Q = Q' = 1$. The product of these automorphisms leads to charges $Q = 0$ and $Q' = 0$, and is in Fock space. The state consists of a soliton moving to the left, having charges $1, 1$ and a soliton moving to the right having charges $-1, -1$. Let $F(t, x)$ be the classical solution of the free wave equation with these properties. Then the corresponding automorphism is implemented in the Fock representation by the unitary operator

$$W(F) = \exp\{i(\phi(\partial_t F) - \pi(F))\}. \tag{7.5}$$

As time goes by, these bumps behave as solitons, in that the shape of the disturbance to the vacuum is unchanged in time, but move with unit speed as outgoing free particles. The state is, however, in Fock space, and has a positive transition probability to (or from) the particle states of the free field. In particular, the two-particles states $|2\rangle$ have a non-zero transition probability to the pair of solitons:

$$P = |\langle 2|W\Psi_0\rangle|^2 > 0. \tag{7.6}$$

According to the theorem of Haag, Doplicher and Roberts [36], in four-dimensional space-time, the superselection group is a compact group. In the present model, however, the charges are real numbers, c say, in number, and appear to be continuous. The states of different values for Q, Q' are inequivalent representations, and so are separated by a superselection rule. The gauge group might then be $(U(1) \times U(1))^c$, which *is* compact. However, in the present book we allow some inequivalent representations of the algebra $\mathcal{A}$ to be linked to each other by weak*-convergence, and so the difference between them is less than in the Haag–Doplicher–Roberts theory.

L. Landau has pointed out that the soliton–antisoliton pair is also non-orthogonal to the one-particle state whose wave function is the sum of the left-moving bump and right-moving bump. In our interpretation, such a one-particle state is unstable in that, with positive probability, it becomes the pair of freely moving solitons, if the apparatus is set up to find these. We know that a decay has occurred when we can detect a positive energy density in two disjoint regions of space which move in opposite directions. We shall see that in $1 + 3$ dimensions, the charge has a minimum size, so that a particle of charge one cannot decay in the same way. This follows from Donaldson's theorem that the cohomology of the bundle is $\mathbf{Z}$, the integers. Thus we might expect to find "elementary" particles of charge $\pm 1, \pm 2, \ldots$. So, together with an electron we might expect to find a double electron of double the mass and double the charge, contrary to experiment. However, a state of charge two, for example, has a non-zero probability of being found as a pair of one-particle states, and so would not be stable; the dynamics would be that of the scattering of two charged particles of equal charge. The same holds for any charge larger than one, or less than -1.

This model has uncountably infinitely many inequivalent representations, and the further study of model does not seem to be so interesting as the model, given below, in four space-time dimensions, which has countably many such representations.

7.2   Instantons, Magnetic Poles and Solitons

In this section we study the case of four space-time dimensions.

Instantons were originally called pseudoparticles [19], but the title "instantons" is now preferred. They are given by explicit solutions to the free Yang–Mills equations in complex projective 4-space with values in the Lie algebra $SU(2)$; they are holomorphic and bounded. By "free" is meant that the curvature and torsion of the connection are zero: it is a flat connection. The original papers also impose either symmetry, which leads to holomorphic solutions, or they impose antisymmetry, which leads to antiholomorphic solutions. We do not require symmetry or antisymmetry, but otherwise we follow Ref. 68.

Let $M = PC^4$ be the complex compact space-time, and let $E \to M$ be an $SU(2)$-bundle over M. We shall use $\wedge$ to denote the exterior product and d for the exterior derivative of the bundle. The gauge group $G = SU(2)$ acts on E, which means that for each $x \in M$, we are given the group element $g(x)$. Let A be a connection one-form on E; recall that under the action of $g \in G$ on E, transforming a frame f as $f \to fg$, the connection one-form transforms as

$$A(fg) = g^{-1}A(f)g + g^{-1}dg. \tag{7.7}$$

Thus, A is a classical non-abelian vector gauge field. The field A is defined only locally. In the overlap of two open sets, there are thus two formulae for A. These differ by a gauge transformation, which means that the gauge-covariant exterior derivative of their difference is zero. The torsion of the connection A is the tensor

$$\kappa(f, h) = dA(f)h + [A(f), A(h)]. \tag{7.8}$$

We want κ, as well as the curvature tensor to be zero; that is, we seek a *flat* connection. Uhlenbeck proved [66] that if a gauge field exists such that $\kappa = 0$ on R^4 and its energy integral is bounded, then it can be extended holomorphically to PC^4 to a solution of the equation $\kappa = 0$. The authors of Ref. 19 constructed an explicit solution for a gauge field A with $\kappa = 0$ when $G = SU(2)$, in two neighbourhoods in PC^4 which between them cover

PC^4, and which in the intersection of the two neighbourhoods differ by a coboundary. The coboundary of A, the gauge field F, is non-zero, and thus provides an example of a non-zero two-cocycle in the holomorphic Čech cohomology of PC^4 with values in $SU(2)$. This work was generalised to give solutions corresponding to two or more instantons and anti-instantons. In the following, we make use of the case of one instanton plus one anti-instanton. When the group is $SU(1)$, rather than $SU(2)$, we show that this case gives a vector that lies in Fock space for the electromagnetic field. For $SU(2)$ and higher, part of the vector lies in Fock space, but the rest involves introducing superselection rules.

We see no reason to try to quantise the Yang–Mills classical field we construct in the next section. This field is not an observable field; it has a charge, so its quantisation would not consist of observables, which are self-adjoint operators. Rather, we use the classical field to provide us with a new representation of the observable fields, E, H, which is inequivalent to the free one: it provides us with *coherent states* not in Fock space.

7.3 An Attempt in $1 + 3$ Dimensions

The electromagnetic field obeys Maxwell's equations

$$\operatorname{div} \boldsymbol{E} = \rho \tag{7.9}$$

$$\operatorname{div} \boldsymbol{B} = 0 \tag{7.10}$$

$$\partial_t \boldsymbol{B} = -\operatorname{curl} \boldsymbol{E} \tag{7.11}$$

$$\partial_t \boldsymbol{E} = \operatorname{curl} \boldsymbol{B} + \boldsymbol{j}. \tag{7.12}$$

The free field arises when ρ and $\boldsymbol{j}$ both vanish. The classical electromagnetic wave is described by a transverse free $\boldsymbol{E}$ and $\boldsymbol{H}$. That is, $\boldsymbol{E}$ and $\boldsymbol{H}$ are mutually orthogonal, of the same magnitude, and both $\boldsymbol{E}$ and $\boldsymbol{H}$ are orthogonal to the momentum of the wave. There are two states, the polarisations, for each momentum. We shall denote an element of this set of solutions by the letter z. The set of such solutions forms a real Hilbert space; let $\langle z_1, z_2 \rangle_{\mathcal{K}}$ be the (real) scalar product.

Following Haag, Doplicher and Roberts [36], we start with the vacuum representation R of the algebra $\mathcal{A}$ of the transverse electromagnetic field, but allow this to be the free field given by the Weyl operators $W_0(z)$. We seek a unit-preserving endomorphism σ of $\mathcal{A}$ so that the representation obtained by the definition

$$R_\sigma(A) := R(\sigma(A)), A \in \mathcal{A} \qquad (7.13)$$

is not equivalent to the representation R. More, we need that at least the space-time automorphisms of $\mathcal{A}$ should be spatial in R_σ, implemented by a unitary operator say $U(a), a \in \mathbf{R}^4$; also that for any one-dimensional time-like translation subgroup with parameter s, the operators $U(s, \mathbf{0})$ should be continuous in s, and that the infinitesimal generator of this group should be bounded below. The dynamics of the field operators, the *transverse* components of $\mathbf{E}$, $\mathbf{H}$, in the representation R_σ, is given by the free automorphism of the free field. However it is not a trivial dynamics in the representation (7.13), and we seek examples such that it is not unitarily equivalent to the free field. For example, this must arise if the longitudinal and time-like components of $\mathbf{E}$, $\mathbf{H}$ obey equations of motion with non-zero ρ or $\mathbf{j}$. The Hamiltonian is not a bounded operator, and neither are the field operators. So these are not in the C*-algebra, and their algebraic properties might not be preserved if we change to an inequivalent representation of the C*-algebra. The commutator with the Hamiltonian gives the time evolution of the field operator. However, the Lie algebra with such commutators might not be preserved under the endomorphism: there might be new terms, an induced interaction. This phenomenon is called anomalous, and it sometimes arises by the appearance of *anomalies* in commutators [1, 24].

First let us start by constructing a class of automorphisms σ_ϕ of $\mathcal{A}$ for the transverse electromagnetic field. Consider a smooth solution $\phi = (\mathbf{E}, \mathbf{H})$ of the free Maxwell equations, which is obtained from a potential A^μ of the form given by Theorem 4. Then by Shale's theorem, Theorem 6, the map $z \mapsto z + \phi$ will not be spatial in R, since ϕ is not square integrable by Theorem 4. We can choose ϕ such that for every $\mathbf{a} \in \mathbf{R}^3$, the function $\phi(\mathbf{x} - \mathbf{a}) - \phi(\mathbf{x})$ *is* square integrable. As a consequence of this, space translations are spatial in R; that is, they are implementable by unitary operators in the representation $W(z) = \sigma_\phi(W_0(z))$, where W_0 is the free representation. To show that σ is invertible, we note that $\sigma_\phi^{-1} = \sigma_{-\phi}$. To show that $\sigma_\phi(1) = 1$, we argue this:

$$1 = W_0(0) = W_0(0z) \mapsto \sigma_\phi(W_0(0z))$$

$$= W_0(0(z + \phi)) = W_0(0) = 1,$$

giving $1 \mapsto 1$ under σ_ϕ. We define σ_ϕ to be given by linearity on the span of the Segal operators; the extension of this by continuity is then linear on the whole of $\mathcal{A}$. To show that $\sigma(XY) = \sigma(X)\sigma(Y)$ for all $X, Y \in \mathcal{A}$, we start by showing the required relation for Segal operators. The action of σ on $W(z)$ is

$$\sigma(W(z)) = e^{iB(\phi,z)}W(z). \tag{7.14}$$

We define the action of σ on finite linear sums of Segal operators to be given by linearity:

$$\sigma\left(\sum_j \lambda_j W(z_j)\right) := \sum_j \lambda_j \sigma(W(z_j)). \tag{7.15}$$

The map σ_ϕ is a *-morphism:

$$\begin{aligned}
\{\sigma_\phi(W(z))\}^* &= \{e^{\{iB(\phi,z)\}}W(z)\}^* \\
&= e^{\{-iB(\phi,z)\}}W^*(z) \\
&= e^{\{iB(\phi,-z)\}}W(-z) \\
&= \sigma_\phi W(-z) \\
&= \sigma_\phi\{W(z)^*\},
\end{aligned}$$

giving the result for Segal operators. The result for finite linear sums then follows immediately. It remains to show that σ_ϕ preserves the product of two elements of the Segal algebra. We find

$$\begin{aligned}
\sigma_\phi(W(z_1)W(z_2)) &= \sigma_\phi\left[\exp\left(-i\frac{1}{2}B(z_1,z_2)\right)W(z_1+z_2)\right] \\
&= \exp\left(-i\frac{1}{2}B(z_1,z_2)\right)\sigma_\phi((W(z_1+z_2))
\end{aligned}$$

From this we see that

$$\begin{aligned}
\sigma_\phi(W(z_1)W(z_2)) &= \exp\left(-i\frac{1}{2}B(z_1,z_2)\right)\exp[iB(\phi,z_1+z_2)]W(z_1+z_2) \\
&= W(z_1)W(z_2)\exp[iB(\phi,z_1)]\exp[iB(\phi,z_2)] \\
&= \sigma_\phi(W(z_1))\sigma_\phi(W(z_2)).
\end{aligned}$$

It follows from this discussion that we can extend σ_ϕ to a *-automorphism of the unique *-algebra generated by the Segal operators by finite sums and products, with compactly supported test functions for the transverse components of $\boldsymbol{E}, \boldsymbol{H}$. Such an automorphism preserves the Slawny norm; for, applied to any element $\alpha_1 W(z_1) + \cdots + \alpha_n W(z_n)$, σ_ϕ is implemented by a unitary operator; the unitary operator that works for a choice of $z_1, \ldots, z_n$ depends on $z_1, \ldots, z_n$. Indeed, let $\mathcal{O} \subset \boldsymbol{R}^3$ be a bounded open set containing the supports of the functions $z_j, j = 1, \ldots, n$, at time $t = 0$. Let $\psi \in \mathcal{K}$ be a solution equal to ϕ in $\mathcal{O}$. Then $\sigma_\psi(W(z_j)) = \sigma_\phi(W(z_j)), j = 1, \ldots, n$. We now show that σ_ψ is implemented by the unitary operator $W(\psi)$. For, putting $W_\psi(z) = W(\psi)W(z)W(\psi)^{-1}$, we get

$$W_\psi(z) = W(\psi + z)\exp\left(\frac{-i}{2}B(\psi, z)\right)W(-\psi)$$

$$= \exp\left(\frac{-i}{2}B(\psi, z)\right)W(z)\exp\left(\frac{-i}{2}B(\psi + z, -\psi)\right),$$

which gives

$$W(\psi)W(z)W(\psi)^{-1} = \exp(-iB(\psi, z))W(z)$$

$$= \sigma_\psi(W(z)).$$

Thus, the map σ_ϕ is continuous in the norm topology, and so can be extended to the completion $\mathcal{A}$ of the *-algebra to give a *-automorphism.

Our intention is to use the representation σ_ϕ, and to keep the free dynamics as the corresponding automorphism group of the C*-algebra of the transverse electromagnetic field. The idea is that the terms ρ and $\boldsymbol{j}$ of the interacting Maxwell equations do not involve the transverse components. Thus for example, the term involving ρ affects the longitudinal component div $\boldsymbol{E}$ only. So the case where $\rho \neq 0$ but is independent of time, and $\boldsymbol{j} = 0$, leads to the same equations for the transverse fields as when $\rho = 0$. Thus in this case the C*-algebra of the transverse fields as well as the time-evolution automorphism, are the same as the case when $\rho = 0$. We now use the fact that the algebra $\mathcal{A}$ is invariant under $\mathcal{P}$. We do a Lorentz transformation on the Maxwell equations with a time-independent ρ, to obtain the same result when ρ does depend on time, and the current is not zero. This is true for all choices of relative velocity for the Lorentz transformation, and for every rotation and space-time translation. Now, the endomorphism of $\mathcal{A}$ generated by the sum $\rho_1 + \rho_2, \boldsymbol{j}_1 + \boldsymbol{j}_2$ is the product of the two automorphisms acting separately. Since neither of these changes $\mathcal{A}$ nor the time-evolution automorphism, neither does the product. Thus, the set of $\rho, \boldsymbol{j}$ with the free transverse algebra $\mathcal{A}$ and evolving under the free automorphism group is a linear space. This space contains densities ρ of any constant velocity, and so the sums of these, getting to a dense set in the space of functions of time. The only thing different is the *representation* of $\mathcal{A}$: the representation obtained from the automorphism σ_ϕ is the coherent representation in which the Segal operator $W(z)$ is represented by

$$\pi_{\sigma_\phi}(W(z)) = W(R_{\sigma_\phi}(z)).$$

This representation is inequivalent to R_0, the Fock representation, as ϕ does not lie in the one-particle space, by Theorem 4.

   This gives us a theory of particle–antiparticle production by a pair of photons, described in Sec. 7.5 of this chapter.

7.4   Flat Connections

A *flat* connection ∇ is a connection for which the curvature tensor is zero, and the torsion tensor is also zero. We shall denote sections of the bundle by symbols $X, Y, Z, \ldots$. The curvature tensor is

$$R^\nabla(X,Y)Z := \nabla_X \nabla_Y Z - \nabla_Y \nabla_X Z - \nabla_{[X,Y]}Z \qquad (7.16)$$

for all X, Y, Z. The torsion tensor of the connection ∇ is

$$T^\nabla(X,Y) = \nabla_X Y - \nabla_Y X - [X,Y]. \qquad (7.17)$$

A manifold with a flat connection ∇, and a real scalar product $\langle\,,\,\rangle$ is called a *statistical manifold* [50] if the covariant derivative of the scalar product is totally symmetric in three variables:

$$\nabla_X \langle Y, Z \rangle = \nabla_Y \langle X, Z \rangle. \qquad (7.18)$$

Such a structure is called a *Codazzi structure* in differential geometry, where the symmetric tensor $\nabla_X \langle Y, Z \rangle$ is called a *cubic form*.

   Given a statistical manifold $(M, \nabla, \langle\,,\,\rangle)$, we define the dual connection ∇^* of ∇ to be given by

$$\nabla_X \langle Y, Z \rangle = \langle \nabla_X^* Y, Z \rangle + \langle Y, \nabla_X Z \rangle, \qquad (7.19)$$

for all X, Y, Z. One can check that $(M, \nabla^*, \langle\,,\,\rangle)$ is also a statistical manifold, called the *statistical dual* of $(M, \nabla, \langle\,,\,\rangle)$.

   Let $\boldsymbol{b}$, $\boldsymbol{h}$ be smooth real vector functions at time zero, such that $\|\boldsymbol{b}\| = \|\boldsymbol{h}\|$ and $\boldsymbol{b} \cdot \boldsymbol{h} = 0$. Consider the representation R_σ of $\mathcal{A}$ obtained by the automorphism σ obtained by adding these to the quantum fields $\boldsymbol{B}$ and $\boldsymbol{H}$ respectively. For the R_σ to be inequivalent to the Fock representation, we must have that the pair fail the square-integrability condition. The

isomorphism giving the smallest non-zero charge can be chosen, as the set
of representations is discrete. For such states of R_σ to have finite electromagnetic energy, the functions e and h must be smooth. For the representation of the little group $SO(2)$ of the Lorentz group, at momentum p
with $p \cdot p = 0$, of charge $n = 1$, we choose $V(\theta) = \exp(2\pi i\theta)$. This means
that for the minimal positive charge e there is one full rotation in the complex plane of the complex field $e + ib$ over the whole space, which we take
to be R^3 at time $t = 0$. The transformed bundle is similar to the *Higgs
bundle*.

Since the addition of e, b to the free field does not alter the commutation
relations, the Segal *-algebra generated by the transformed fields is the same
as when e, b are zero. Since the *-algebra has a unique C*-norm, by Slawny's
theorem, it follows that the C*-algebra, obtained as the norm-completion
of the Segal *-algebra of the transformed field, is the same as that of the
free field; but since e, b are not square integrable, the representation of the
C*-algebra is not equivalent to that given by the free field.

7.5   Pair Production from Photon–Photon Scattering

In the previous section, the case when $n = 1$ gives rise to a stable particle,
known as an instanton or soliton. The momenta of the state might be
anything; suppose that it is in the positive x-direction, and at time zero,
the place in R^3 might be chosen to be near the origin. Now consider the
case when this transformation is followed by one with $n = -1$, near the
origin at time zero, but with momentum in the $-x$ direction. The combined
transformation has $n = 0$, and so the representation of the combined field is
the Fock representation of the free field. The vector describing the state is
an exponential vector, with a test function in the one-particle space which
is the sum of the two parts, and this is not orthogonal to the two-particle
space of the photon. However, as we watch the state of free photons move
in time, the field with $n = +1$ moves to the right and that with $n = -1$
moves to the left, and for large times the solitons separate and appear
as a soliton–antisoliton pair. The convergence is in the weak* topology in
Fock space. Thus, a pair of photons which collide near $x = 0$, moving in
opposite directions, has a non-zero scalar product with the state of the
soliton–antisoliton pair: the probability of pair production is positive.

The transition probability from the pair of photons to the state of the
soliton–antisoliton pair can be found by using the Uhlmann calculus [67].

7.6 Non-Abelian Gauge Fields

We now generalise the construction of the representation of the electro-magnetic field to a reducible one by allowing the addition to the free-field solution $F^0_{\mu\nu}$ to be matrix functions, say $\mathcal{F}_{\mu\nu}$, of space at time zero, rather than just real-valued fields. Thus, at time zero we try a representation of the Maxwell tensor $F_{\mu\nu}$ of the form

$$F_{\mu\nu}(\boldsymbol{x}) = F^0_{\mu\nu}(\boldsymbol{x}) \otimes I + 1 \otimes \mathcal{F}_{\mu\nu}(\boldsymbol{x}). \qquad (7.20)$$

For example, we may take I to be the 2×2 identity matrix, and $\mathcal{F}_{\mu\nu}(\boldsymbol{x})$ to be antisymmetric 2×2 matrix functions of $\boldsymbol{x}$ which all commute with each other at all values of their positions $\boldsymbol{x}$. We also require the imaginary exponentials of a corresponding vector potential around any closed loop to be multiples of the identity. The same remarks hold for any dimension instead of 2; the following discussion is given for dimension 2.

Again we construct the observable C*-algebra by using the free fields $\mathcal{E}$ and $\mathcal{H}$ at time zero, smeared with smooth functions on $\boldsymbol{R}^3$ of compact support. The time evolution is again taken to be free dynamics.

We construct (by a choice of gauge) a vector potential A_μ for the transverse part of the quantised field. With this choice of gauge, we add to $\boldsymbol{E}(\boldsymbol{x})$ a 2×2 vector-valued matrix function $\boldsymbol{E}_0$ of $\boldsymbol{x}$. We add a 2×2 vector-valued matrix function $\boldsymbol{H}_0$ to $\boldsymbol{H}(\boldsymbol{x})$, such that the transverse components of the tensor $\mathcal{F}_{\mu\nu}$ are zero. Each of these matrix functions is taken to be the sum of two terms; one with Fourier transform zero unless $k_x > 0$, and the other with Fourier transform zero unless $k_x < 0$. Of course, any other direction than that in the x-direction could have been chosen. The part with $k_x > 0$ is chosen so that its part proportional to I_2 is a soliton, and the corresponding part with $k_x < 0$ is a corresponding anti-soliton. The sum has a part proportional to the identity matrix I_2, which lies in $\mathcal{K}_1 \otimes I_2$. The factor $\mathcal{K}_1$ is the one-particle space. The time evolution of this part is given by the free time-evolution τ_t of $\boldsymbol{E}, \boldsymbol{H}$, and for large positive times this has two parts, one whose support in x is large-positive and for the other, the support in x is large-negative. We can follow the left-moving part of the soliton for a time t in any bounded open set $\mathcal{O} \in \boldsymbol{R}^3$ until the right-moving part of the soliton has completely left $\mathcal{O}$, except for a part of small norm. From then on, the time evolution is the same as for the free soliton, except for a small correction, which can be made smaller and smaller by choosing t larger and larger. Let σ_t be the dynamics of the free soliton. It follows that the time evolution of $\mathcal{A}(\mathcal{O})$ becomes σ_t, in that for all $A \in \mathcal{A}(\mathcal{O})$, we have

$$(\tau_t^{-1}\sigma_t)A \to A \qquad (7.21)$$

as $t \to \infty$. This is true for any bounded set $\mathcal{O}$, and so holds for any A in the union over bounded $\mathcal{O}$. It then follows easily that it holds for any $A \in \mathcal{A}$. Thus, we have shown that $\tau_t^{-1}\sigma_t \to 1$ as $t \to \infty$ in the weak* topology.

The representation of the C*-algebra obtained by using the endomorphism, got by adding matrix functions of x to each component of $E(x)$ and $H(x)$ at time zero, is reducible. Since the state of a suitable soliton–antisoliton pair converges as time goes to infinity to the coherent state of the free quantised photon field, we can compute the probability that at large times it contains say two photons. This is the Uhlmann transition probability between the two states [67].

The convergence is in the weak* topology, not the strong; this is because strong convergence would imply that the Fock representation contains some soliton states, which is not true. If we form the direct sum of the two representations, the Fock representation and the representation containing the soliton, then we expect to get at least weak convergence.

Chapter 8

Conclusion

In this book, we have briefly studied quasifree quantum fields, and argued that they could provide non-zero scattering for some particles. This is surprising, since by definition, a quasifree quantum field is one with zero as the value of any connected time-ordered product of n quantised fields. We start with the Wightman functions of the free tranverse electromagnetic fields, and construct the C*-algebra $\mathcal{A}$ of the Fock representation. We note that the algebra $\mathcal{A}$ has some non-Fock representations. Some states in the Fock representation converge as time goes to infinity to states which are not in Fock space, but can be interpreted as products of a specific non-Fock pair of a particle and its antiparticle. The convergence is in the weak* topology, and the states are normalised positive linear forms on our C*-algebra $\mathcal{A}$. At infinite time, these lie in an inequivalent representation, possibly reducible. For all finite times, the free photon system is in a state which has a non-zero scalar product with some reducible exponential states on our algebra; these can be represented as an infinite sum of vacuum, one-photon states, two-photon states, and so on. Each n-photon state lies in Fock space, but the norm of the sum is infinite. The theory thus gives a prediction, for example, of the probability that two photons will produce the compound state being considered: it is the square of the two-photon contribution to the compound. The interesting fact is that there is no freedom to choose the effective coupling constant: it is uniquely determined by the theory. If we make the conjecture, that there do not exist Wightman theories that are not quasifree in four space-time dimensions, then it would follow that all types of particles might be produced by examining such models.

The theory of representations of systems of fields obeying the Haag–Kastler axioms yields a system of superselection rules, based on groups of the form of $SU(1)$, $SU(2)$, or $SU(3)$ etc. This is the triumphant conclusion of Haag, Doplicher, Roberts and coworkers. These authors assume that the field should be an interacting one (which they do not possess yet). However, the same conclusions also follow if the quantised field is free; and also it could be quasifree. I suggest that the quasifree electromagnetic field might be enough to produce all the particles, with their couplings predicted correctly.

Bibliography

[1] Adler S., *Phys. Rev.* **177**, 2426, 1969.

[2] Albeverio S., Gottschalk H. and Wu J.-L., Models of Local Relativistic Quantum Fields with Indefinite Metric (in All Dimensions), *Commun. Math. Phys.* **148**, 509–531, 1997.

[3] Albeverio S. and Hoegh-Krohn R., Euclidean Markov Fields and Relativistic Quantum Fields from Stochastic Partial Differential Equations in Four Dimensions, *Phys. Lett.* B **177**, 175–179, 1980.

[4] Albeverio S. and Hoegh-Krohn R., Construction of Interacting Local Relativistic Quantum Fields in Four Space-Time Dimensions, submitted to *Phys. Lett. B*, 1987.

[5] Ali S. T., Antoine J.-P. and Gazeau J.-P., *Coherent States, Wavelets and Their Generalizations*, Springer-Verlag, 2000.

[6] Al-Rashed M. H. A. and Zegarlinski B., Noncommutative Orlicz Spaces Associated to a State, *Studia Mathematica* **180**, 199–209, 2007.

[7] Araki H., Relative Entropy for States of von Neumann Algebras, *Publ. RIMS, Kyoto University* **11**, 809–833, 1976.

[8] Araki H., *Mathematical Theory of Quantum Fields*, Oxford University Press, 1999.

[9] Ashtekar A., Geometric Issues in Quantum Gravity, in Ref. 39, pp. 173–194.

[10] Asselmeyer T., Generation of Source Terms in General Relativity by Differential Structures, *Classical and Quantum Gravity* **14**, 749–758, 1996.

[11] Asselmeyer T., Exotic R^4 and Quantum Field Theory, in Ali C. B. E. (ed.), *International Conference on Quantum Theory and Symmetries*, IOP Publishing, Bristol, 2011.

[12] Asselmeyer T., ArXiv:1112.4885, 2011.

[13] Asselmeyer T. and Król J., Constructing a Quantum Field Theory from Space-Time, ArXiv:1107.3458, 2011.

[14] Asselmeyer T. and Rosé H., Dark Energy and 3-Manifold Topology, *Acta Phys. Polonica* **38**, 3633–3639, 2007.

[15] Bargmann V., On a Hilbert Space of Analytic Functions and an Associated Integral Transform, *Commun. Pur. Appl. Math.* **14**, 187–214, 1961.

[16] Bargmann V., Note on Wigner's Theorem on Symmetry Operations, *J. Math. Phys.* **5**, 862–868, 1964.

[17] Bargmann V., Wightman A. S. and Wigner E. P., Princeton Lecture Notes, unpublished.

[18] Baston R. J. and Eastwood M. G., *The Penrose Transform*, Oxford Science Publications, 1989.

[19] Belavin A. A., Polykov A. M., Schwarz A. S. and Tyupkin Yu. S., Pseudoparticle Solutions of the Yang–Mills Equations, *Phys. Lett.* B **59**, 85–87, 1975.

[20] Bell J. S. and Jackiw R., *Nuovo Cimento* **51**, 47, 1969.

[21] Borchers H. J., Energy and Momentum as Observables in Quantum Theory, *Commun. Math. Phys.* **2**, 49–54, 1966.

[22] Bose S. N., Planck's Gesetz und Lichtquanten Hypothesis, *Z. Physik* **26**, 178, 1924.

[23] Bratteli O. and Robinson D. W., *Operator Algebras and Quantum Statistical Mechanics*, 2nd ed. Springer-Verlag, New York, 1979, Theorem 5.2.8.

[24] Buchholz D. and Jacobi P., On the Nuclearity Condition for Massless Fields, *Lett. Math. Phys.* **13**, 313–323, 1987.

[25] Buchholz D. and Wichmann E. H., Causal Independence and the Energy-Level Density of States in Quantum Field Theory, *Commun. Math. Phys.* **106**, 321–344, 1986.

[26] Ciolli F., Net Cohomology and Completeness of Superselection Sectors, *Rev. Math. Phys.* **21**, 735–780, 2009.

[27] Corrigan E. and Fairlie D. B., Scalar Field Theory and Exact Solutions to a Classical SU(2)-Gauge Theory, *Phys. Lett.* B **67**, 69–71, 1977.

[28] Cronin J. W., Recent Experimental Developments in Weak Interactions, in Hagen C. R., Guralnik G. and Mathur V. S. (eds.), *Proceedings of 1967 International Conference on Particles and Fields*, Interscience, New York, London, Sydney, pp. 3–20, 1967.

[29] Donaldson S. K., Self-Dual Connections and the Topology of Smooth Manifolds, *Bulletin AMS* **8**, 81–83, 1983.

[30] Doplicher S. and Roberts J. E., Why there is a Field Algebra with a Compact Gauge Group Describing the Superselection Structure in Particle Physics, *Commun. Math. Phys.* **131**, 51–107, 1990.

[31] Einstein A., Sitz Berichte, *Preuss. Akademie Berlin* **22**, 261, 1924.

[32] Frobenius F. G., see Curtis C. W., *Pioneers of Representation Theory: Frobenius, Burnside, Schur and Brauer*, Providence R. I., 1999.

[33] Gajdzinski C. and Streater R. F., Anomaly Meltdown, *J. Math. Phys.* **32**, 1981–3, 1991.

[34] Glimm J. and Jaffe A. M., *Quantum Physics: A Functional Point of View*, 2nd ed., Springer-Verlag, Berlin, 1987.

[35] Gross L., Norm Invariance of Mass Zero Equations under the Conformal Group, *J. Math. Phys.* **5**, 687–695, 1964.

[36] Haag R., *Local Quantum Physics*, 2nd ed., Springer-Verlag, Berlin, Heidelberg, New York, 1992.

[37] Haag R. and Kastler D., An Algebraic Approach to Quantum Field Theory, *J. Math. Phys.* **5**, 848–861, 1964.

[38] Higgs P. W., Broken Symmetries and the Masses of Gauge Bosons, *Phys. Rev. Lett.* **13**, 508–509, 1964.

[39] Huggett S. A., Mason L. J., Tod K. P., Tsou S. T. and Woodhouse N. M. J., *The Geometric Universe*, Oxford University Press, Oxford, 1998.

[40] Jackiw R., Nohl C. and Rebbi C., Conformal Properties of Pseudoparticle Configurations, *Phys. Rev. D* **15**, 1642–1646, 1977.

[41] Jost R., *General Theory of Quantized Fields*, Amer. Math. Soc., 1965.

[42] Kibble T. W. B., *J. Math. Phys.* **9**, 315–324, 1968.

[43] Klauder J. R. and Skagerstam B.-S., *Coherent States: Applications in Physics and Astrophysics*, World Scientific, Singapore, 1985.

[44] Landau L., private communication.

[45] Landsman N. P. and Wiedermann U. A., Massless Particles, Electromagnetism and Rieffel Induction, *Rep. Math. Phys.* **7**, 923–958, 1995.

[46] Lee T. D. and Yang C. N., *Phys. Rev.* **104**, 254, 1956.

[47] Leyland P. and Roberts J. E., The Cohomology of Nets over Minkowski Space, *Commun. Math. Phys.* **62**, 173–189, 1978.

[48] Mackey G., *Induced Representations of Groups and Quantum Mechanics*, W. A. Benjamin, 1968.

[49] Mackey G., Quantum Mechanics from the Point of View of Group Representations, in Flato M., Sally P. and Zuckerman, G. (eds.), *Applications of Group Theory in Physics*, Amer. Math. Soc., pp. 210–254, 1985.

[50] Matsuzoe H., Computational Geometry from the Viewpoint of Affine Differential Geometry, *Lecture Notes in Computer Science* **5416**, 103–123, 2009.

[51] Mickelsson J., Hilbert Space Cocycles as Representations of $(3+1)$-D Current Algebras, *Lett. Math. Phys.* **28**, 97–106, 1993.

[52] Nelson E., The Free Markov Field, *J. Funct. Anal.* **12**, 211–227, 1973.

[53] Penrose R., Solutions of the Zero Rest-Mass Equations, *J. Math. Phys.* **10**, 38–39, 1969.

[54] Segal I. E., Tensor Algebras over Hilbert Space, I, *Trans. Amer. Math. Soc.* **81**, 196–134, 1956.

[55] Segal I. E., *Mathematical Problems of Relativistic Physics*, Amer. Math. Soc., 1963.

[56] Segal I. E., Complex-Wave Representation of the Free Boson Field, in Gohberg I. and Kac, M. (eds.), *Topics in Functional Analysis*, Math. Supplement Studies, Academic Press, 1978.

[57] Shale D., Linear Symmetries of Free Boson Fields, *Trans. Amer. Math. Soc.* **103**, 149–167, 1962.

[58] Simon B., *The* $P(\varphi)_2$ *Euclidean (Quantum) Field Theory*, Princeton Series in Physics, Princeton University Press, 1974.

[59] Slawny J., On Factor Representations and the C^*-Algebra of the CCR, *Commun. Math. Phys.* **24**, 151–170, 1972.

[60] Streater R. F., Symmetry Groups and Non-Abelian Cohomology, in Tanner E. A. and Wilson R. (eds.), *Noncompact Lie Groups and Some of their Applications*, pp. 333–340, 1994.

[61] Streater R. F., *Lost Causes in and beyond Physics*, Springer, pp. 148–153, 2007.

[62] Streater R. F., A Theory of Scattering Based on Free Fields, preprint dated 16 October, 2009, and published in Hara T., Matsui T. and Horoshima F. (eds.), *Mathematical Quantum Field Theory and Renormalisation Theory*, Math-for-industry, Vol. 30, Kyushu University, pp. 10–20, 2011. See also ArXiv/hep-th:1001.1882.

[63] Streater R. F. and Wightman A. S., *PCT, Spin and Statistics, and All That*, Princeton University Press, 2000.

[64] Streater R. F. and Wilde I. F., Fermion States of a Boson Field, *Nucl. Phys. B* **24**, 561–575, 1970.

[65] Symanzik K., Euclidean Quantum Field Theory, I: Equations for a Scalar Model, *J. Math. Phys.* **7**, 510–525, 1966.

[66] Uhlenbeck K., Removable Singularities in Yang–Mills Fields, *Commun. Math. Phys.* **83**, 11–29, 1982.

[67] Uhlmann A., The "Transition Probability" in the State Space of a *-Algebra, *Rep. Math. Phys.* **9**, 273–279, 1976.

[68] Ward R. S. and Wells R. O., *Twistor Geometry and Field Theory*, Cambridge University Press, 1990.

[69] Wightman A. S., Quantum Field Theory in Terms of Vacuum Expectation Values, *Phys. Rev.* **101**, 860, 1956.

[70] Wigner E. P., *Atomic Physics and the Theory of Spectral Multiplicity*, Princeton University Press, 1935.

[71] Wigner E. P., Unitary Representations of the Inhomogeneous Lorentz Group, *Ann. Math.* **40**, 149–204, 1939.

[72] Witten E., Some Exact Multi-Instanton Solutions of Classical Yang–Mills Theory, *Phys. Rev. Lett.* **38**, 121–124, 1977.

[73] Witten E., Introduction to Cohomological Field Theories, *Int. J. Mod. Phys. A* **6**, 2273, 1991.

[74] Wu C. S., Ambler E., Hayward R. W., Hoppes D. D. and Hudson R. F., *Phys. Rev.* **105**, 1413, 1957.

[75] Yau T. H., The Connection between an Euclidean Gauss Markov Vector Field and the Real Proca Wightman Field, *Commun. Math. Phys.* **41**, 267–271, 1975.

[76] Yau T. H., Construction of Quantum Fields from Euclidean Tensor Fields, *J. Math. Phys.* **17**, 241, 1976.

Index